Current Topics in Microbiology 251 and Immunology

Springer-Verlag Berlin Heidelberg GmbH

Lymphoid Organogenesis

Proceedings of the Workshop held at the
Basel Institute for Immunology
5th - 6th November 1999

Edited by F. Melchers

With 62 Figures and 7 Tables

Springer

Fritz Melchers, Ph.D.
Basel Institute for Immunology
Postfach
Grenzacherstr. 487
CH-4005 Basel
Switzerland

Cover illustration:
The cover shows a thymus epithelial cell reaggregate (RFTOC) following implantation under the kidney capsule. Thymus epithelial cell reaggregates were generated from ~200 nl of epithelial cell suspensions in vitro. Reaggregates had a diameter of ~1 mm after 2-3 days in culture. From 4 weeks after transplantation onwards, histological analysis revealed that RFTOC resembled a lymphoid tissue (left side) clearly demarcated from the host kidney (right side). Such epithelial reaggregate grafts provide a functional environment for host bone marrow-derived T cell development. For details see the article by H.-R. Rodewald, this volume.

Cover design:
Hanspeter Stahlberger, Basel Institute for Immunology, Switzerland

Editorial assistance:
Leslie Nicklin, Basel Institute for Immunology, Switzerland

Cover:
Design & production GmbH, Heidelberg, Germany

ISSN 0070-217X
ISBN 978-3-642-63186-3 ISBN 978-3-642-57276-0 (eBook)
DOI 10.1007/978-3-642-57276-0

Originally published by Springer-Verlag Berlin Heidelberg New York in 2000
Softcover reprint of the hardcover 1st edition 2000
Library of Congress Catalog Number 15-12910

Typesetting: Camera-ready by authors
Printed on acid-free paper SPIN: 10770607 27/3020 hu - 5 4 3 2 1 0

Preface

Lymphoid organs are sites of lymphocyte production and lymphocyte responses. The organs are interconnected by blood and lymph through which lymphocytes migrate. Several cell lineages contribute to the structure of lymphoid organs. Hematopoietic stem cells are the pluripotent progenitors of erythroid, myeloid and lymphoid cells. Hemangioblasts are thought to be the common progenitors of hematopoietic and vascular endothelial cells, the latter giving rise to the vessels in which hematopoietic and lymphoid cells travel. Mesenchyme, deriving from the cephalic region of the neural crest, develops the network of fibroblasts and stromal cells which is an important part of the supportive environment for the development and responses of lymphocytes. Mesenchyme also is the origin of bone-forming osteoblasts. Finally, epithelial cells which develop from the third pharyngeal pouch have the ability, e.g. in the thymus, to shape the repertoire of the developing T lymphocytes. As they develop and build the organ, the different lineages of cells are known to interact with each other, and to help each other in their development.

For decades lymphoid organs were dissected experimentally into single cells in order to identify the different stages of development of the different lineages, and to understand their function. Now that we have a better understanding of these cellular pathways and functions, we are facing a new challenge: to reconstruct and assemble that which we have taken apart.

It has been clear for a long time that after transplantation of a lymphoid organ, hematopoietic stem cells can regenerate the compartments of the organ, provided that the rest of its architecture - the stroma, the epithelia and the vessels - is intact. Ahead lies the even greater challenge to assemble also these other architectural elements of a lymphoid organ by transplanting stem cells. The workshop on lymphoid organogenesis was convened to review current knowledge of and experimental skills involved in this grand project to build a lymphoid organ from its individual cellular components.

Together with my two colleagues and co-organizers, Hans-Reimer Rodewald and Antonius Rolink, I thank the participants not only for their excellent presentations and lively Basel-style discussions but in particular for contributing to this volume. I thank too the students of the Basel Institute – Claudia Waskow, Axel Bouchon, Maurus Curti, John Gatfield, Jan Kisielow, Hinchi Kong, Anya Langenkamp, Dieter Naeher, Christoph Schaniel, Thomas Seidl, and Greg Terszowski – for projecting the slides; Hans Spalinger for the photography;

Michel Delétraz, Michel Dekany, Peter Menini and Dorothy Poms for the catering, and, last but not least, Leslie Nicklin for organizing the meeting in Basel and for her never-ceasing efforts and help in editing the papers in this volume. Finally I thank Mme. France Benguerel, Lausanne, for her generous donation to the workshop, and F. Hoffmann-La Roche Ltd., the company which fully supports the Basel Institute for Immunology, and thus supported this workshop.

FRITZ MELCHERS

Table of Contents

II Vessel Development

III Thymus Development

IV Lymphoid Tissue Development

List of Contributors

(Their adresses can be found at the beginning of their respective chapters.)

I

Stem Cells – Hematopoietic, Epithelial, Others

Human Mesenchymal Stem Cells: Progenitor Cells for Cartilage, Bone, Fat and Stroma

M. F. Pittenger, J. D. Mosca and K. R. McIntosh
Osiris Therapeutics, Inc., 2001 Aliceanna Street, Baltimore, MD 21231, USA

Introduction

Bone marrow provides the rich milieu necessary to maintain myeloid and lymphoid progenitor cells throughout the life of an organism. At least two stem cell populations have been identified in marrow, the hematopoietic stem cell (HSC) and the mesenchymal stem cell (MSC). The HSC has been characterized in many ways, but much remains to be learned about its intrinsic potential and interactions with other cells of the marrow environment. We have studied the human stem cell population for mesenchymal tissues that resides in adult bone marrow. These MSCs potentially have the ability to differentiate to all mesenchymal cell types, including osteocytic, chondrocytic, adipocytic, myocytic, tenocytic, and also dermal and stromal lineages (1, 2). We have sought to understand the potential role(s) that MSCs play in healthy individuals and their response to trauma, disease or aging.

Such studies require standardized cell populations and methods of assaying the cell's potential. Cells grown from a single human cell from marrow differentiated to multiple lineages (chondrocytes, adipocytes and osteoblasts) demonstrating their status as human mesenchymal stem cells (hMSCs) (3). This chapter will briefly describe the isolation, culture and characterization of the hMSCs from adult bone marrow, their differentiation potential, and highlight some of their surface molecules that predict their interaction with lymphoid lineages. The cells from human marrow are the best characterized of MSCs from mammalian species, and we and our colleagues have isolated these cells from marrow samples of more than 500 volunteer donors (3). The cultivation of hMSCs permits an enhanced understanding of this important progenitor cell for multiple tissue types and the potential to develop new therapeutic strategies for tissue regeneration including the regeneration of damaged stroma in individuals undergoing treatment with chemotherapy or radiation (4).

The formation of bone following the transplantation of bone marrow to ectopic sites stimulated interest in identifying the source of osteoprogenitor cells from marrow (5, 6). Current research on marrow-derived adult osteogenic precursor cells traces its origins to the pioneering work of Alexander Friedenstein and colleagues beginning in the 1960s. Their in vitro cultivation of cells from guinea pig bone marrow produced cells that appeared fibroblastic, would form colonies (hence the name fibroblast colony forming cell, FCFC) and were osteogenic when placed in diffusion chambers and implanted subcutaneously (7, 8, 9).

Multiple groups have subsequently investigated marrow-derived progenitor cells for connective tissues or hematopoietic support. Studies of marrow-derived progenitor cells have been carried out in mouse, rat, rabbit, dog and man. The majority of these reports have been concerned with the differentiation to a single lineage although several authors investigated multiple lineages (10-15). Recently, a study from Prockop and colleagues showed that isolated marrow cells could engraft into multiple tissues including bone, cartilage, marrow, spleen and lung when infused into irradiated host mice (16). The cells isolated from marrow that attach and expand in culture have been given various names including marrow stromal fibroblasts (MSFs), bone marrow mesenchymal cells (BMMCs) marrow stromal cells (MSCs), mesenchymal stem cells (MSCs), marrow progenitor cells (MPCs), colony forming unit-fibroblastic (CFU-F) (2, 9, 11, 21, 27). Whether these populations from different laboratories represent, in general, the same cell type(s) is not yet resolved.

In vitro studies to directly address the interaction between marrow stromal cells and those of the hematopoietic system by Dexter and colleagues showed that the marrow environment contributed to the support of hematopoietic stem cells (17). Further research on marrow-derived stromal cells and lymphopoietic precursors has been done in the mouse, and useful stromal cell lines have been derived (18-21). Many of the growth factors and cytokines necessary to maintain human HSCs in vitro have been identified and include interleukin (IL)-3, IL-6, IL-7, erythropoietin (epo), stem cell factor (SCF), granulocyte-colony stimulating factor (G-CSF), and granulocyte-macrophage-colony stimulating factor (GM-CSF) (19). Purified sources of these molecules are available. However, the optimal formulation for HSC growth and maintenance of the stem cell phenotype has not yet been described and HSCs continue to be grown on feeder layers of cells to provide a stromal environment. The hMSCs have been shown to provide such stromal support to HSCs (32).

While the mouse would seem the ideal system to test the ability of MSCs to support hematopoiesis, the isolation of homogeneous primary mouse MSCs has proven difficult. However, certain mouse species have been reported to yield good populations of cells for study (22). We routinely work with MSCs from several species but focus on the human cells with which we have the most experience.

Human MSCs from Adult Bone Marrow

Haynesworth and Caplan reproducibly isolated a population of marrow-derived cells that formed bone and cartilage when seeded onto a ceramic carrier and implanted subcutaneously into athymic mice (1). This in vivo assay was used in an iterative fashion for the selection of lots of fetal bovine serum that support the expansion of hMSCs (23). Experience indicates that only selected lots of fetal bovine serum are useful for hMSC expansion and until a generally acceptable chemically defined, serum-free medium is developed, the need for serum screening will continue (23). The question of whether bone and cartilage seen in diffusion chambers or ceramic implants arose from separate cell populations or from a common progenitor cell remained.

The development of in vitro assays allowed us to test independent differentiation to each lineage (3). The results showed the presence of only the desired differentiated phenotype. This differs from reports of differentiating ectopic cultures of mammalian embryonic stem (ES) cells where a variety of differentiated phenotypes are seen. While there is much to be learned about the stem cell nature of the hMSCs from adult bone marrow and their ability to give rise to the different mature mesenchymal phenotypes, they do provide the rich starting material for such work. Experience shows that hMSCs respond to growth factors and cytokines as well as to their culture environment, including spatial organization, cell density and basal nutrients.

The procedure to isolate multipotential hMSCs from bone marrow has been improved and published in detail (1, 3, 24). Briefly, a marrow sample is drawn from the iliac crest of the donor. Marrow is withdrawn using a bone marrow aspiration needle connected to a syringe containing heparin. The marrow sample is washed with saline and the cell suspension is layered onto a 25 ml density cushion of 1.073g/ml. Samples are subjected to centrifugation. The light density upper layer, including cells at the interface, are collected, washed, and the cells collected by centrifugation. The nucleated cells are resuspended in Dulbecco's modified Eagles medium (DMEM) with 10% FBS from selected lots. About 30% of the original nucleated cell number is present in the light density fraction and only perhaps 0.01 to 0.1% of these are actually hMSCs. The contaminating cells are mostly of hematopoietic origin, do not attach, and are removed with sub-sequent medium changes. The cells are plated and incubated at 37° C with 5% CO_2 with medium changes twice a week. Initial colony formation is evident in 3-4 days and the well spread hMSCs are ready to subculture in 12-16 days. Thereafter the cells are subcultured 1:3 about once a week. The cells continue to grow well but do not expand indefinitely. We generally use the passage 1 – 4 hMSCs although karyotype analysis as late as passage 12 was normal (3).

In vitro Differentiation of hMSCS

Methods have been developed to test the multilineage potential of cultured hMSCs to the chondrogenic (25-27), adipogenic (4) and osteogenic lineages (28,

29) and these methods work with several species (see Fig. 1). Briefly, chondrogenic differentiation of hMSCs is assayed in a micromass pellet culture by centrifuging 250,000 hMSCs in a polypropylene tube in a serum-free Dulbecco's modified Eagles medium (DMEM) containing 10 ng/ml TGF-β3, 100 nM dexamethasone, 50 μg/ml ascorbic acid 2-phosphate, 100 μg/ml sodium pyruvate, 40 μg/ml proline and a commercially available insulin-transferrin-selenium solution (ITS^+, Collaborative Biomedical, Bedford, MA, USA).

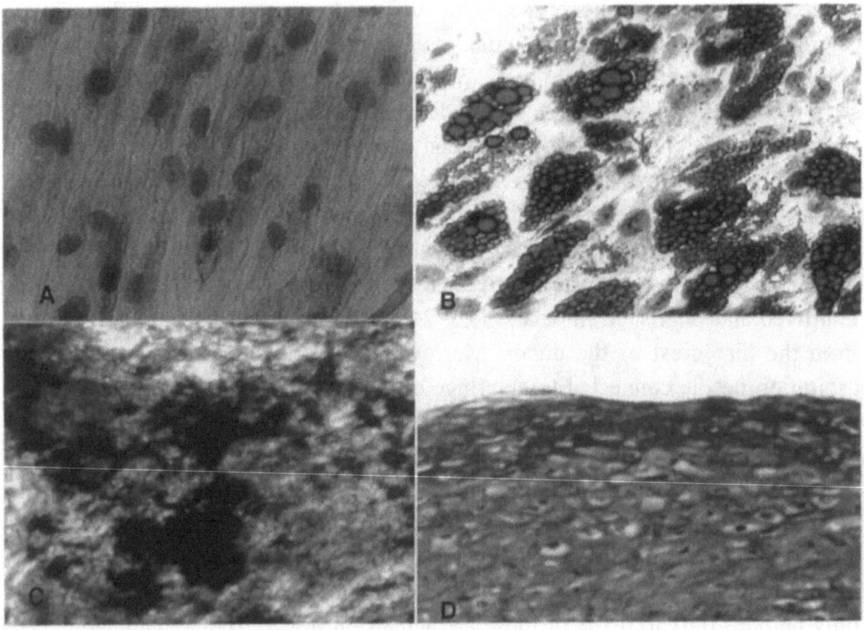

Fig. 1. Differentiation of hMSCs to the adipogenic, osteogenic and chondrogenic lineages. The hMSCs grow as contact inhibited monolayers for many passages (A) and can be directed to differentiate to different mesenchymal cell types including adipocytes (B) osteoblasts (C) or chondrocytes (D) under appropriate conditions. A-Bright field image of hematoxylin stained hMSCs, B-Bright field image oil red O and hematoxylin counterstained hMSCs showing accumulation of lipid vacuoles following adipogenic differentiation. C-Bright field image of alkaline phosphatase and von Kossa silver stained hMSCs following osteogenic differentiation. Note the dark silver stained calcium deposits. D-Brightfield image of section through a micromass pellet culture of hMSCs undergoing chondrogenic differentiation stained for collagen II expression. Typical chondrocyte lacunae are evident.

The cells do not attach to the culture tube and instead form a ball of cells. Chondrogenic differentiation occurs over the next 2-3 weeks with the expression of collagen type II, aggrecan, other extracellular proteoglycans, etc.

The hMSCs can be induced to become adipogenic in vitro by treating confluent monolayer cultures with 0.5 mM isobutyl methylxanthine, 1 μM dexamethasone, 100 μM indomethacin and 10 ug/ml insulin (MDII) in DMEM with 10% FBS. Best results were seen with three 48 hr treatments of MDII separated by 24 hrs in DMEM with 10% FBS. The adipogenic differentiation is apparent visually by the accumulation of lipid containing vacuoles which can be stained with Nile red or oil red O. The adipogenic cells upregulate the expression of peroxisome proliferation-activated receptor (PPARγ), lipoprotein lipase (LPL), C/EBPα, fatty acid binding protein aP2, and leptin (3). Differentiation is evident after the first treatment but more cells commit to this lineage in subsequent treatments.

The osteogenic differentiation of hMSCs occurs in sub-confluent monolayer cultures of hMSCs in DMEM with 10% FBS, 100 nM dexamethasone, 10 mM β-glycerolphosphate and 50 μM ascorbic acid 2-phosphate (28, 29). Differentiation results in the increase in alkaline phosphatase activity, upregulation of osteocalcin, osteopontin, collagen type I and calcium deposition over 1-2 weeks.

These differentiation protocols should allow a detailed investigation of the molecular pathways responsible for committing the hMSCs to each of these lineages. The use of the hMSCs for therapeutic intervention is currently under further investigation.

Non-antigen Interactions of hMSCs with HSCs and Lymphocytes

Numerous studies have indicated that bone marrow stroma provides a favorable microenvironment for hematopoietic stem cells (HSCs) (reviewed in 19). Culture-expanded MSCs produce a number of growth factors and cytokines including G-CSF, SCF, LIF, M-CSF, IL-6 and IL-11 (30). Additional ligands detected by RT-PCR include IL-7, IL-8, IL-12, IL-14, and IL-15 while IL-2, IL-3, IL-4, IL-10 and IL-13 have not been detected (3, 32). hMSCs have been shown to perform a stromal cell function as demonstrated by their ability to maintain CD34+ stem cells in long-term cultures (32). Other reports suggest that MSCs enhance the ability of HSCs to self-replicate and differentiate into various lineages and similar results have been obtained using MSC-derived adipocytes (31, 33, 40, Mbalaviele and Pittenger, unpublished). In regard to the immune system, hMSCs appear to be necessary for B cell lymphopoiesis as these cells are not normally generated from CD34+ HSCs in long term culture, but have been seen when hMSCs were included in the cultures. hMSC support of B cell lymphopoiesis could not be replaced by a cytokine cocktail in contrast to the generation of myeloid cells that could be produced from HSCs with cytokines alone (Mbalaviele and Mosca, personal communication).

Cell-cell interactions between T cells and thymic stromal cells have been studied by a number of investigators, in part due to the interest in stromal cells as mediators of T cell selection (35-38). hMSCs are also of interest due to their potential role in extrathymic differentiation of T cells (34). In rosetting experiments with hMSCs and autologous T lymphocytes (with both cell types in suspension) or cell binding assays (MSCs bound to wells), minimal binding between resting T cells and MSCs was observed. However, activation of T cells with either antigen or mitogens greatly enhanced binding to MSCs. Flow cytometry analysis of the T cell subpopulations demonstrated that both CD4+ and CD8+ T cells bound equally well to the hMSCs (Mosca and McIntosh, in preparation). Evaluation of hMSC surface markers has demonstrated that these cells constitutively expressed a number of adhesion molecules including ICAM-1, ICAM-2, VCAM-1, L-selectin, CD72, and LFA-3 (3). Since T cells express counter-receptors to many of these molecules, it is tempting to speculate that hMSCs or their progeny (including stromal cells) may be involved in physiologic events involved in T cell maturation in either bone marrow or thymus as described above. In this regard, it is of interest that human MSCs injected into sheep fetuses can be found months later in the thymus, that appear positive for CD73 (A. Flake, personal communication). These data from several independent lines of research suggest that hMSCs promote lymphopoiesis in marrow stroma and perhaps in thymic stroma.

Antigen-related Interactions Between MSCs and T Cells

The surface molecules of culture-expanded hMSCs suggest that these cells may be capable of presenting antigens to T cells. Recent experiments show that hMSCs express MHC Class I molecules but little if any MHC Class II (McIntosh and Mosca, unpublished). However, exposure of hMSCs to the pro-inflammatory cytokine IFN-γ up-regulated Class I expression and induced the expression of moderate amounts of Class II on the cell surface. There was no expression of the co-stimulatory molecule, B7-2, on MSCs as evaluated by flow cytometry, nor was there detectable mRNA for this molecule as assessed by RT-PCR. B7 expression was not enhanced by treating the cells with pro-inflammatory cytokines. Cell surface expression of ICAM-1, another costimulatory molecule, was detectable by flow cytometry and this molecule was increased by IL-1α, TNF-α, and IFN-γ.

The ability of hMSCs to express both Class I and Class II molecules under inflammatory conditions indicated that these cells should be capable of presenting antigens to both Class II restricted CD4+ T cells and Class I restricted CD8+ cells. The outcome of antigen presentation depends on whether the MSCs can co-stimulate the T cell. Sufficient levels of co-stimulation would be expected to trigger IL-2 production resulting in activation and lymphoproliferation, whereas T cell receptor triggering in the absence of co-stimulation would be expected to result in T cell anergy (39). We have produced T cell lines specific for tetanus toxoid from peripheral blood cells obtained from several donors and produced hMSCs from these same donors. When the hMSCs were incubated with IFN-γ (100 U/ml) for 3 days and pulsed with tetanus toxoid, they were able to induce low, but significant, proliferative responses from the T cell lines (McIntosh and Mosca, unpublished). Treatment with IFN-γ was essential to stimulate T cell proliferation. Retroviral transduction of either B7-1 or B7-2 into the hMSCs enhanced the T cell proliferative response to a level that was comparable to professional antigen presenting cells from peripheral blood. It will be of interest to determine whether antigen-pulsed hMSCs can activate resting T cells or whether this interaction will result in the induction of an unresponsive state. Therefore, hMSCs may have a therapeutic use for inducing tolerance to renegade immune responses in autoimmune disease or in preventing transplant rejection.

Acknowledgements

We would like to thank our colleagues Daniel R. Marshak, Robert J. Deans and Christian van den Bos for discussion and careful reading of the manuscript. Part of this work was supported by the Defense Advanced Research Projects Agency.

References

1. Haynesworth SE, Goshima J, Goldberg VM, Caplan AI (1992) Characterization of cells with osteogenic potential from human bone marrow. Bone 13:81-88.
2. Caplan, AI (1991) Mesenchymal stem cells. J. Orthop. Res. 9:641-650.
3. Pittenger MF, Mackay AM, Beck S, Jaiswal R, Douglas R, Mosca J, Moorman MA, Simonetti D, Craig S, Marshak DR (1999) Multilineage potential of human mesenchymal stem cells. Science 284:143-147.
4. Lazarus HM, Haynesworth SE, Gerson SL, Rosenthal NS, Caplan AI (1995) Ex vivo expansion and subsequent infusion of human bone marrow-derived stromal progenitor cells (mesenchymal progenitor cells): Implications for therapeutic use. Bone Marrow Transpl. 16:557-564
5. Urist, MR, McLean FC (1952) Osteogenic potency and new bone formation by induction in transplants to the anterior chamber of the eye. J. Bone Joint. Surgery Am. Vol. 34:443-470.
6. Tavassoli M, Crosby W (1968) Transplant of marrow to extramedullary sites. Science 161:548-556.
7. Friedenstein, AJ, Piatetzky-Shapiro, II, Petrokova, KV (1966) Osteogenesis in transplants of bone marrow cells. J. Embyrol. Exp. Morphol.16:381-390.
8. Friedenstein AJ, Petrakova KV, Kurolesova AI, Frolova GP (1968) Heterotopic transplants of bone marrow: Analysis of precursor cells for osteogenic and haematopoietic tissues. Transplantation 6:230-47.

9. Friedenstein, AJ (1973) Determined and inducible osteogenic precursor cells. In: K. Elliott and D. Fitzsimmons (ed.s) Hard Tissue Growth, Repair and Remineralization. Ciba Found. Sympos.11:169-185.
10. Ashton BA, Allen TD, Howlett CR, Eaglesom CC, Hattori A, Owen M (1980) Formation of bone and cartilage by marrow stromal cells in diffusion chambers in vivo. Clin. Orth. Rel. Res. 151:294-307.
11. Friedenstein, AJ, Chailakhyan RK, Gerasimov UV (1987) Bone marrow osteogenic stem cells: in vitro cultivation and transplantation in diffusion chambers. Cell Tissue Kinet. 20:263-272.
12. Gregoriadis, AE, Heersche JNM, Aubin JE (1988) Differentiation of muscle, fat, cartilage, and bone from progenitor cells present in a bone derived clonal cell population: effect of dexamethasone. J. Cell Biology 106:2139-2151.
13. Beresford JN, Bennett JH, Devlin C, LeBoy PS, Owen M (1992) Evidence for an inverse relationship between the differentiation of adipocytic and osteogenic cells in rat marrow stromal cell cultures. J. Cell Sci. 102:341-351.
14. Poliard A, Nifuji A, Lamblin D Forest C, Kellermann O (1995) Controlled conversion of an immortalized mesodermal progenitor cell towards osteogenic, chondrogenic and adipogenic pathways. J. Cell Biol. 130:1461-1472.
15. Dennis JE, Merriam A, Awadalla A, Yoo JU, Johnstone B, Caplan AI (1999) A quadripotent mesenchymal progenitor cell isolated from the marrow of an adult mouse. J. Bone Miner. Res. 14:700-709.
16. Pereira RF, Halford KW, O'Hara MD, Leeper DB, Sokolov BP, Pollard MD, Bagasra O, Prockop DJ (1995) Cultured adherent cells from marrow can serve as long-lasting precursor cells for bone, cartilage and lung in irradiated mice. Proc. Natl. Acad. Sci. USA. 92:4857-4861.
17. Dexter, TM, TD Allen, LG Lajtha (1977) Conditions controlling the proliferation of haematopoietic stem cells in vitro. J. Cell Physiology 91:335-344.
18. Zipori, D, A Friedman, M Tamir, D. Silverberg, Z. Malik (1984) Cultured mouse marrow lines: interactions between fibroblastoid cells and monocytes. J. Cellular Physiol. 118:143-152.
19. Dorshkind K (1990) Regulation of hemopoiesis by bone marrow stromal cells and their products. Ann. Rev. Immunol. 8:111-137.
20. Pietrangeli, CE, Hayashi SI, Kincade PW (1988) Stromal cell lines which support lmphocyte growth: characterization, sensitivity to radiation and responsiveness to growth factors. Eur. J. Immunol. 18:863-872.
21. Gimble, JM, Dorheim MA, Cheng Q, Medina K, Wang CS, Jones R, Koren E, Pietrangeli M, Kincade PW (1990) Adipogenesis in murine bone marrow stromal cell line capable of supporting B lineage lymphocyte growth and proliferation. Eur. J. Immunol. 20:379-387.
22. Phinney DG, Kopen G, Isaacson RL, Prockop D (1999) Plastic adherent stromal cells from the bone marrow of commonly used strains of inbred mice: variations in yield, growth and differentiation. J. Cell. Biochem. 72:570-585.
23. Lennon DP, Haynesworth SE, Bruder SP, Jaiswal N, Caplan, AI (1996) Development of a Serum Screen for Mesenchymal Progenitor Cells from Bone Marrow. In Vitro Animal Cellular & Developmental Biology 32: 602-611.
24. M.F. Pittenger, G. Mbalaviele, J. Mosca, M. Black, D.R. Marshak (2000) Adult Mesenchymal Stem Cells in Primary Mesenchymal Cells MR Koller, BO Palsson and JRW Masters, Editors. Dordrecht, Kluwer Academic Publishers. (In Press).
25. Barry FP, Johnstone B, Pittenger MF, Mackay AM, Murphy JM (1997) Modulation of the chondrogenic potential of human bone marrow-derived mesenchymal stem cells by TGFβ1 and TGFβ3. Trans. Ortho Res. Soc. 22:228.
26. Mackay AM, Beck SC, Murphy JM, Barry FP, Chichester CO, Pittenger MF (1998) Chondrogenic differentiation of cultured human mesenchymal stem cells from marrow. Tissue Engin. 4:415-428.
27. Yoo JU, Barthel TS, Nishimura K, Solchaga L, Caplan AI, Goldberg VM, Johnstone B (1998) The chondrogenic potential of human bone-marrow derived mesenchymal progenitor cells. J. Bone Joint Surg. 80A:1745-1757.
28. Jaiswal N, Haynesworth SE, Caplan AI, Bruder SP (1997) Osteogenic differentiation of purified, culture-expanded human mesenchymal stem cells in vitro. J. Cell. Biochem. 64: 295-312.

29. Bruder SP, Jaiswal N, Haynesworth SE (1997) Growth kinetics, self-renewal and the osteogenic potential of purified human mesenchymal stem cells during extensive subcultivation and following cryopreservation. Journal of Cellular Biochem. 64:278-294.
30. Haynesworth SE, Baber MA, Caplan AI (1996) Cytokine expression by human marrow-derived mesenchymal progenitor cells in vitro: effects of dex and IL-1α. J. Cell. Physiol. 166: 585-592.
31. Allay JA, Dennis JE, Haynesworth SE, Majumdar MK, Clapp DW, Schultz LD, Caplan AI (1997) LacZ and interleukin-3 expression in vivo after retroviral transduction of marrow-derived human osteogenic mesenchymal progenitors. Hum Gene Therapy 8:1417-1427.
32. Majumdar MK, Thiede MA, Mosca JD, Moorman M, Gerson SL (1998) Phenotypic and functional comparison of marrow-derived mesenchymal stem cells and stromal cells. J. Cell. Phys. 176:57-66.
33. Mbalaviele G, Lui L, Lee K, Novelli EM, Buyaner D, Deans R, Civin C, Mosca JD. Human mesenchymal stem cells can enhance human CD34+ cell repopulating NOD/SCID mice. Submitted for publication.
34. Barda-Saad M, Rozenszajn LA, Ashush H, Shav-Tal Y, Ben Nun A, Zipori D (1999) Adhesion molecules involved in the interactions between early T cells and mesenchymal bone marrow stromal cells. Exp. Hematol. 27:834-844.
35. Hojo H, Yaguchi M, Nagasu M, Aizawa S, Nakano M, Harigaya K, Handa H, Toyama K (1994) Establishment of a human thymic stromal cell line (R-3-4) and its adhesive capacity with T cells in vitro. Hematol. Pathol.8:177-185.
36. Fort MM, Pardoll DM (1996) Can bone marrow-derived thymic stromal cells mediate the positive selection of class I-restricted T cells? Cell. Immunol. 171:74-79.
37. Nonoyama S, Nakayama M, Abe J, Kohsaka T, Kobayashi N, Yata J (1992) Characterization of thymic lymphocytes forming rosettes with stromal cells. Tohoku J. Exp. Med. 168:467-474.
38. Owen JJ, Moore NC (1995) Thymic-stromal-cell interactions and T-cell selection. Immunol. Today 16:336-338.
39. Mueller DL, Jenkins MK, Schwartz RH (1989) Clonal expansion versus functional clonal inactivation: A costimulatory signalling pathway determines the outcome of T cell antigen receptor occupancy. Ann. Rev. Immunol. 7:445-480.
40. Cheng L, Qasba P, Padmavathy V, Thiede M (2000) Human mesenchymal stem cells support megakaryocyte and pro-platelet formation from $CD34^+$ hematopoietic progenitor cells. J Cellular. Physiol. In press.

Investigating the Genetic Control of Stem Cell Behavior

R.L. Phillips
Department of Molecular Biology, Princeton University, Princeton, New Jersey 08544 USA

The precise control of hematopoietic stem cell fate decisions involves molecules that are differentially expressed in the stem cell versus its non-stem progeny. The construction of representative and high quality cDNA libraries containing stem cell-specific sequences is a first step in elucidating stem cell control mechanisms. Automated bioinformatics in conjunction with high-throughput random sequencing and high-density parallel array hybridization studies make it possible to dissect stem cell molecular pathways and networks. It is the differential analysis of these interacting pathways which will provide the greatest insights into the biological differences underlying stem cell behavior.

Introduction

Hematopoietic stem cells sit atop the blood cell developmental hierarchy and are responsible for the production of at least eight separate lineages of mature blood cells, including red cells, cells of the granulocytic, monocytic, and lymphoid series, and the platelet-producing megakaryocytes [1]. Stem cells are operationally defined by their behavior and exhibit two important properties. First they must be totipotent with respect to the hematopoietic system; that is, able to produce all of the blood cell lineages [2]. They are unique among the members of the hematopoietic hierarchy for their second property—the ability to undergo extensive self-renewal (i.e., increase their own numbers) without lineage commitment [3,4]. Unlike the rest of the cells in the bone marrow, in which division is accompanied by progressive restriction of cell lineage choices [5], stem cells seem to be able to uncouple differentiation from cell division. The mechanism that controls this decision, whether to self-renew or proceed down a differentiation pathway, is not known and has proven difficult to study. While it is possible using genetic approaches to identify molecules in which the disruption ("knockout") of both alleles results in hematopoietic insufficiency or overproduction [6-8], these studies and others like them only focus on the end-point of hematopoietic differentiation and not on the earliest stages, at which the critical self-renewal decision takes place.

There is evidence to suggest that the regulation of this important stem cell decision is not entirely stochastic but is under genetic control [9,10]. Since the stem cells are different from their non-stem progeny with respect to self-renewal potential [11], it is likely that the genes regulating this potential are differentially expressed in the stem cell relative to the very immature non-stem progenitors. In other words, some element of the control mechanism for stem cell expansion must be expressed at different levels in cells that do not share this potential. In order to isolate likely control molecules from the thousands of expressed genes in stem cells, it is necessary to examine gene expression on as wide a scale as possible (ideally genome-wide) and to focus on those genes that are specifically expressed in the stem cells. A new era in functional genomics begins with the genetic profiling of functionally-defined cell populations.

The advent of high-throughput screening techniques has made it possible to rapidly determine the genetic profile of well-defined populations of cells and to compare this profile with those of other cell populations. By purifying large enough numbers of stem cells and non-stem progenitors, it is possible to construct representative cDNA libraries that accurately reflect the gene expression of the population. A comparison of gene expression profiles among populations of cells that have been selectively enriched or depleted of stem cell activity will aid in the identification of differentially expressed genes and point the way toward a biological mechanism for differential cell fate control in the hematopoietic system.

Use of High-Density Arrays in Stem Cell Genomics

Investigations into stem cell molecular biology are made more complete through the use of arrays of hundreds or thousands of cDNAs (as either plasmids or PCR products) on standard nylon membranes or on small glass chips. These systems are still being developed and are quite expensive, but they allow the parallel interrogation of entire cDNA libraries. Gene expression in the various hematopoietic compartments may be compared directly through differential hybridization. Several manufacturers now produce non-redundant arrays of up to 18,000 murine or human genes on a single hybridization substrate.

One of the immediate problems that face researchers using high-density arrays is the issue of sensitivity. Genes, especially early-acting regulatory genes, may be expressed at low levels that are below the limits of detection of hybridization. As with any hybridization there is a desirable level of non-specific background (the "noise") that must be distinguished from the signal. Having no background at all makes it difficult to accurately determine the position of the individual closely-packed dots on the arrays. The reduction of background and the enhancement of sensitivity (increase of the signal-to-noise ratio) may be achieved through a molecular subtraction of the probes before hybridization. As depicted in Figure 1, subtraction of the cDNAs found in the second probe from the first and vice versa tends to remove co-hybridizing non-specific cDNAs and increase the differential hybridization ratio of the two signals. Use of this technique "highlights" the expression differences, which can be further investigated by a more quantitative technique such as RT-PCR.

The use of ready-made, non-redundant commercial arrays of human or mouse clones has several advantages over custom arrays. The non-redundancy itself is an advantage in efficiency, as arrays made from clones selected at random will have a significant number of duplicated "sister" clones. Arraying a normalized cDNA library would tend to cut down on—but not eliminate—the duplicate clone problem. Another advantage is that the ready-made arrays contain clones that have already been sequenced. This means it is possible to go from array hybridization to differential sequence analysis in a day's time. Lastly, since commercial arrays are standardized and are used by a large number of investigators, results are easily compared between experiments or even between biological systems. The main disadvantage of commercial arrays is that they do not contain clones that are completely novel; that is, that have never before been sequenced. This problem will be overcome shortly with advances in the human genome project. Soon arrays (or sets of arrays) will be available that contain all open reading frames in a given genome,

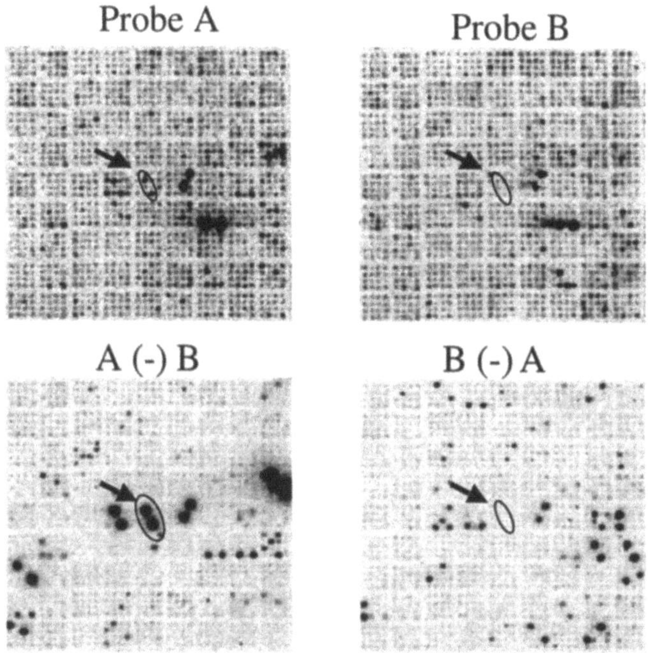

Fig 1. Subtracted probes magnify expression differences. Identical high-density arrays were hybridized with whole cDNA probes (top) or bidirectionally subtracted cDNA probes (bottom). For the clone pair shown at the arrow (and circled), the ratio of hybridization intensities between populations A and B were measured at 7.5 (top) and 149 (bottom). The use of a pre-subtraction step increases the signal-to-noise ratio significantly.

and the simultaneous monitoring of gene expression can take place on a scale that is truly genome-wide. Array technology becomes more and more useful as the genome sequencing projects near their completion.

Analysis of Differentially Expressed Sequences

High-throughput screening techniques result in huge amounts of data in a relatively short period of time. The use of a database allows for the automated analysis of the cDNA sequences obtained in the screen and provides clues for answering more complex biological questions. We have placed our high-throughput data into a database on the World Wide Web. The Stem Cell Database is a repository for sequence data from high-throughput random sequencing of stem cell libraries as well as high-density array screens. Because all the sequences are kept in the computer, they can be automatically compared (using BLAST) to those in the public sequence databases. As shown in Figure 2, we routinely compare nucleotide data and conceptual translations to Genbank nonredundant (nucleotide and protein) databases, the Swissprot protein database, dbEST Expressed Sequence Tags (ESTs), and the EST contigs in the DOTS project at the University of Pennsylvania (www.cbil.upenn.edu/DOTS). These comparisons identify individual clones to

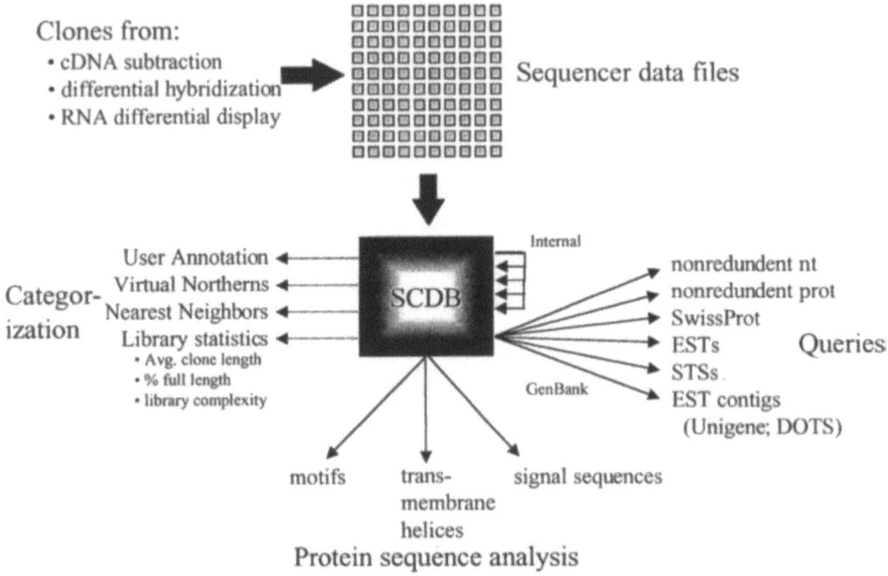

Fig. 2. Automated bioinformatics. Clones from high-throughput screens or random sequencing efforts are compared to those in publicly available nucleotide and protein sequence databases. Potential open reading frames are scanned for motifs, transmembrane helices, or signal peptides. The large amount of medium-quality data for each clone is turned into a high-quality "executive summary" by a human annotator. Efforts are underway to computerize this labor-intensive but very important final stage of annotation.

allow their categorization in the database with respect to function. In addition, new members of particular protein families may be identified by their extensive homology with known protein domains. These data taken together provide an idea of what is happening in the stem cell specifically.

How can one isolate stem cell-specific molecules when the stem cells themselves can not be purified to homogeneity? We are fortunate to have several different protocols for the purification of stem cells from murine fetal liver and adult bone marrow as well as human marrow. Each of these also identifies closely related cell populations that are negative for stem cell properties. The fetal liver sort protocol of Lemischka [12] identifies stem cells as Sca-1^{pos}c-Kit^{pos}AA4.$1^{pos}$$Lin^{neg/lo}$ and non-stem cells as AA4.1^{neg}. The murine bone marrow protocol of Nakauchi [13] reveals stem cells as Sca-1^{pos}c-$Kit^{pos}$$CD34^{neg/lo}$$Lin^{neg}$ cells. Spangrude [14] has made use of a mitochondrial vital dye, rhodamine-123 (Rh-123), to define them in bone marrow as Thy-1^{lo}Sca-1^{pos}c-Kit^{pos}Rh-$123^{lo}$$Lin^{neg}$. Bartelmez [15] has used this dye as well as the DNA stain Hoechst 33342 (Ho) to identify stem cells in the Ho^{lo}Rh-$123^{lo}$$Lin^{neg}$ fraction of bone marrow.

Sequences that appear in subtracted libraries from more than one stem cell source (or ideally, all stem cell sources) but are absent from all of the non-stem cell sources are prime candidate stem cell-specific molecules (Figure 3). While each protocol results in populations that are highly enriched for stem cells, they also contain different types of non-stem co-purified cells. The exact makeup of these co-purified cells depends on the protocol being used. An analysis of the overlap among large sequence sets derived from multiple stem cell sources can reduce the signals from the non-stem cells. Additionally, the relative extent of overlap between

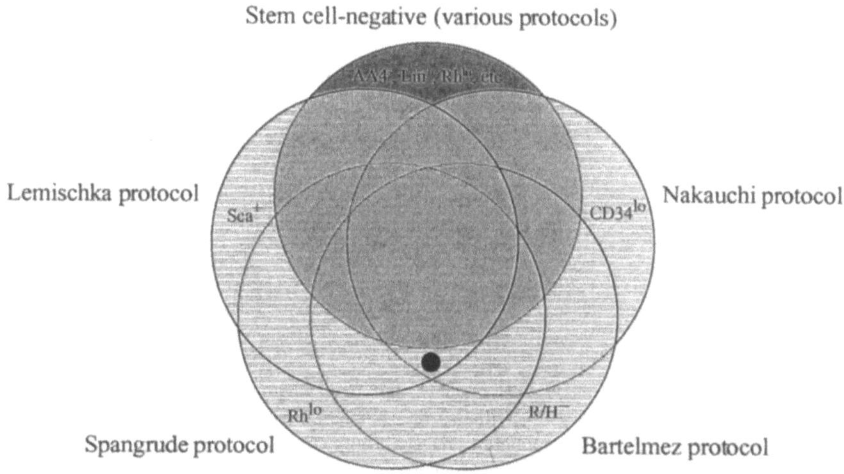

Fig. 3. Use of high-density arrays to isolate stem cell-specific molecules. Shown is a Venn diagram of sequences that are assumed to be specific to cell populations derived from each protocol. The star designates the much smaller population of candidate stem cell-specific molecules that can be evaluated using more expensive and laborious techniques. Each of the murine stem cell sort protocols shown here enriches for stem cells to a similar degree but generates different types of non-stem co-purified cells. Thus, even though stem cells can not be purified to homogeneity, their genetic program may be uncovered.

sequence sets is a direct measurement of the relatedness of the underlying cell populations. Given these data for several subpopulations of the hematopoietic hierarchy it should be possible to re-evaluate the accepted hierarchy in light of the molecular similarities among the subpopulations. In much the same way as molecular evolution analysis has forced (in some cases) a re-drawing of the evolutionary tree, the redefinition of hematopoietic compartments as large arrays of expressed genes can allow a more accurate depiction of the blood cell developmental hierarchy.

One beneficial by-product of a complete survey of stem cell-specific molecules is the identification of surface antigens that can be used for positive or negative selection of stem cells from bone marrow. We have already demonstrated the utility of high-throughput screens in this regard by identifying and characterizing CD27 expression on stem cell-containing populations [16]. CD27 was shown to be highly differentially expressed in multiple high-throughput screens, and this expression pattern was confirmed by subsequent flow cytometry studies.

Integration of Data

As more and more information is uncovered about each individual sequence, it becomes apparent that some of the sequences relate to one another through commonalities such as membership in a particular developmental pathway or interaction with the same known protein. Insights into the biology underlying such

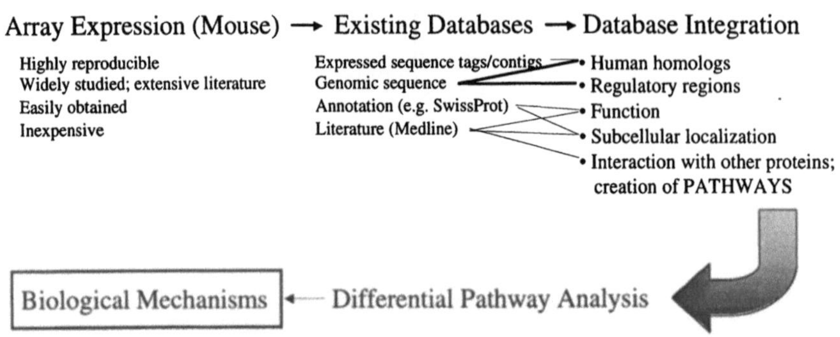

Fig. 4. Expression studies can provide insights into mechanisms. The placement of high-throughput information into databases facilitates the annotation process, and integration of the experimental database with other genomic databases allows for a differential analysis of developmental pathways that are activated or repressed in the target cell population.

large sets of sequences will come when the analysis data and bioinformatics are all taken together in an investigation of integrated pathways and networks within the target cells (Figure 4 and [17]). The mass of information emerging from high-throughput projects can readily be placed into a computerized database that will store and categorize data from sequence analysis. This database can in turn be integrated with public genomics databases to provide information about homologous human genes (and potential regulatory regions based on mouse/human homologies [18]), potential subcellular localization and putative function, and, most importantly, the complex interactions between differentially expressed genes in known developmental and regulatory pathways. All of this is leading to the ability (with the help of the computer to keep track of the huge amount of information simultaneously) to examine regulatory mechanisms in light of the activation or repression state of important pathways.

The study of stem cell biology is about to enter its final and most important phase, where we will be able to pinpoint the mechanisms that regulate the most interesting stem cell behaviors. A deeper understanding of stem cell biology, made possible by genome-wide analyses and world-wide database integration, will allow for the manipulation of stem cells in vitro and in vivo and will lead to a realization of their promise for therapeutics.

Acknowledgements

The work described here would not have been possible without the dedication of Ihor Lemischka and the members of his laboratory. I thank G. Christian Overton and especially Brian Brunk for their help in getting the Stem Cell Database put together and making it into a useful resource. The work described here was supported by the U.S. National Institutes of Health, the American Cancer Society, and Imclone Systems, Inc. (New York).

References

1. Metcalf D, Nicola NA, Robb L (1997) Differentiation commitment in normal hemopoiesis and leukemic transformation. J Cell Physiol 173: 131-134.
2. Morrison S, Uchida N, Weissman IL (1995) The biology of hematopoietic stem cells. Annu Rev Cell Dev Biol 11: 35-71
3. Potten CS, Loeffler M (1990) Stem cells: attributes, cycles, spirals, pitfalls and uncertainties. Lessons for and from the crypt. Development 110: 1001-1020.
4. Jordan CT, Lemischka IR (1991) Clonal and systemic analysis of long-term hematopoiesis in the mouse. Genes Dev 4: 220-232.
5. Suda J, Suda T, Ogawa M (1984) Analysis of differentiation of mouse hemopoietic stem cells in culture by sequential replating of paired progenitors. Blood 64: 393-399.
6. McKercher SR, Torbett BE, Anderson KL, Henkel GW, Vestal DJ, Baribault H, Klemsz M, Feeney AJ, Wu GE, Paige CJ, Maki RA (1996) Targeted disruption of the PU.1 gene results in multiple hematopoietic abnormalities. EMBO J 15: 5647-5658.
7. Pandolfi PP, Roth ME, Karis A, Leonard MW, Dzierzak E, Grosveld FG, Engel JD, Lindenbaum MH (1995) Targeted disruption of the GATA3 gene causes severe abnormalities in the nervous system and in fetal liver haematopoiesis. Nat Genet 11: 40-44.
8. Robb L, Elwood NJ, Elefanty AG, Kontgen F, Li R, Barnett LD, Begley CG (1996) The scl gene product is required for the generation of all hematopoietic lineages in the adult mouse. EMBO J 15: 4123-4129.
9. Phillips RL, Reinhart AJ, Van Zant G (1992) Genetic control of murine hematopoietic stem cell pool sizes and cycling kinetics. Proc Natl Acad Sci U S A 89: 11607-11611.
10. Muller-Sieburg CE, Riblet R (1996) Genetic control of the frequency of hematopoietic stem cells in mice: mapping of a candidate locus to chromosome 1. J Exp Med 183: 1141-1150.
11. Osawa M, Nakamura K, Nishi N, Takahasi N, Tokuomoto Y, Inoue H, Nakauchi H (1996) In vivo self-renewal of c-Kit+ Sca-1+ Lin(low/-) hemopoietic stem cells. J Immunol 156: 3207-3214.
12. Moore KA, Ema H, Lemischka IR (1997) In vitro maintenance of highly purified, transplantable hematopoietic stem cells. Blood 89: 4337-4347.
13. Osawa M, Hanada K, Hamada H, Nakauchi H (1996) Long-term lymphohematopoietic reconstitution by a single CD34-low/negative hematopoietic stem cell. Science 273: 242-245.
14. Spangrude GJ, Johnson GR (1990) Resting and activated subsets of mouse multipotent hematopoietic stem cells. Proc Natl Acad Sci U S A 87: 7433-7437.
15. Wolf NS, Kone A, Priestley GV, Bartelmez SH (1993) In vivo and in vitro characterization of long-term repopulating primitive hematopoietic cells isolated by sequential Hoechst 33342-rhodamine 123 FACS selection. Exp Hematol 21: 614-622.
16. Wiesmann A, Phillips RL, Mojica M, Pierce LJ, Searles AE, Spangrude GJ, Lemischka I (2000) Expression of CD27 on murine hematopoietic stem and progenitor cells. Immunity 12: 193-199.
17. Tamayo P, Slonim D, Mesirov J, Zhu Q, Kitareewan S, Dmitrovsky E, Lander ES, Golub TR (1999) Interpreting patterns of gene expression with self-organizing maps: methods and application to hematopoietic differentiation. Proc Natl Acad Sci U S A 96: 2907-2912.
18. Stojanovic N, Florea L, Riemer C, Gumucio D, Slightom J, Goodman M, Miller W, Hardison R (1999) Comparison of five methods for finding conserved sequences in multiple alignments of gene regulatory regions. Nucleic Acids Res 27: 3899-3910.

Precursor B Cells from *Pax-5*-deficient mice – Stem Cells for Macrophages, Granulocytes, Osteoclasts, Dendritic Cells, Natural Killer Cells, Thymocytes and T Cells

A.G. Rolink and F. Melchers
Basel Institute for Immunology, Grenzacherstrasse 487, CH-4005 Basel, Switzerland

Introduction

At least three major cell differentiation lineages cooperate to build the system of lymphoid organs, each one derived from a separate stem cell. Hematopoietic stem cells give rise to T-lymphoid cells in the thymus and to B-lymphoid and myeloid cells in the bone marrow. Fibroblastic and epithelial stem cells seed the bone marrow and thymus to provide the stromal cell environment in bone marrow, and the cortical and medullary epithelial cell layers in the thymus. Interactions between the fibroblasts and epithelial cells on the one side and the hematopoietic cells on the other induce both lineages to their differentiation pathways, and organize different regions in the primary lymphoid organs – fetal liver, thymus, bone marrow – as much as they do so in the secondary lymphoid organs, such as spleen and lymph nodes. Finally, endothelial cells form the vessels through and between lymphoid organs which allow hematopoietic cells in a process of attachment and transmigration to enter the vessels, migrate in them from one site through blood and lymph, and exit into another site. The formation of bone needs yet another cell lineage to generate osteoblasts which build bone (Rodan and Harada 1997), in balance with the osteolytic osteoclasts which are provided by hematopoietic stem cells.

Hematopoietic Stem Cells

Hematopoietic stem cells are expected to possess at least four properties (Spangrude et al. 1988; Morrison and Weissman 1994; Morrison et al. 1994; Morrison et al. 1997).

1) Upon division at least one daughter cell retains the state of a stem cell (self-renewal);
2) The other daughter cell has a (variable) tendency to differentiate along several possible pathways, in several distinct stages of cellular development;

3) Upon transplantation into a suitably receptive host, stem cells can home to their proper sites in the body (unless implanted at these sites), and can retain the properties of self-renewal and pluripotency (long-term reconstitution);
4) Secondary and subsequent transplantations retain the properties of self-renewal, pluripotency and long-term reconstitution.

Stem cells of the three lineages – hematopoietic, fibroblastic, epithelial and endothelial – first arise during embryonic development. For two of the three lineages – hematopoietic and endothelial vascular – a common progenitor, the hemangioblast, has been implicated (Eichmann, this volume, see further references there). It is a continuously debated issue whether stem cells persist, in a dormant state, throughout life. For the self-renewing, pleuripotent, long-term reconstituting hematopoietic stem cell this appears to be the case. Hence, the isolation and possible *in vitro* propagation of such stem cells appears an easier, for the human case also ethically defendable task whenever they can be found in adult tissues.

In order to successfully attempt tissue reconstitution from a stem cell, the recipient host has to allow this stem cell to seed the proper sites – in the case of hematopoietic stem cells the bone marrow. The transplanted stem cells have to be able to compete for these sites with endogenous stem cells which may already occupy them. In the case of hematopoietic cell development it is practically mandatory to irradiate the host, hence destroy some progenitor compartments in order to allow hematopoietic regeneration from the transplanted stem cells. Even then, the differentiation along a certain pathway of myeloid or lymphoid lineage in the host is greatly enhanced whenever the host is, in fact, defective in that lineage, i.e. misses the differentiated cell compartment before stem cell transplantation. One well-studied example is the set of severely combined immunodeficient mouse strains which are defective in parts of the Ig and TcR gene rearrangement machinery (*SCID, RAG-1*$^{-/-}$ and *2*$^{-/-}$) and, hence, are blocked in their lymphoid cell developments at the early precursor cell states in the primary lymphoid organs.

Although, in principle, one single stem cell may be sufficient to repopulate all compartments of differentiated cells of a lineage this may, in practice, be a difficult task, not least because stem cells isolated *ex vivo* may have a tendency to differentiate and, thereby, lose their stem cell properties while they find themselves in a foreign environment, either *in vitro* or during transplantation *in vivo* until they can reach their proper sites, i.e. their proper environment. A large number of stem cells is definitely needed for the *in vitro* introduction of genes, either by transfection of DNA containing the gene, or by infection with a suitable viral vector containing the gene. Since they are rare at any time of development, the isolation and purification of a large enough number of stem cells is a considerable experimental problem. Consequently, stem cells should be able to grow in the right tissue culture conditions without losing their self-renewal, pluripotency and long-term reconstituting capacities whenever genetic alterations such as the ones intended in gene therapy experiments are planned. Until recently such stem cells were not available for either of the three lineages involved in lymphoid organogenesis.

Pax-5-deficient preB Cells Possess Stem Cell-like Activity

Surprisingly, self-renewing, pluripotent, long-term reconstituting, *in vitro* expandable hematopoietic cells have recently been found in a mouse in which the homeobox gene *Pax-5* has been made non-functional by targeted inactivation (Urbanek et al. 1994; Nutt et al. 1999, Rolink et al. 1999). These cells, which are found in the bone marrow, have only one defect so far identified in hematopoiesis: they cannot make B cells since they are blocked in the development of that lineage at the transition of preB-I to large preB-II cells (Urbanek et al. 1994; Nutt et al. 1997). In fact, even more surprisingly, the cells with these stem cell properties have the phenotype of a preB-I cell. Like preB-I cells from wildtype mice these *Pax-5*$^{-/-}$ preB-I cells carry D_H J_H rearrangements on both IgH chain alleles (but not $V_HD_HJ_H$ rearrangements, nor any V_LJ_L rearrangements at either the κL or λL chain gene locus). They express B-lineage-specific genes such as VpreB and λ5, which form the surrogate L chain (Nutt et al. 1997; Nutt et al. 1998). They express the preB cell receptor and B cell receptor anchoring proteins, Igβ and Igα, the latter to a lesser extent. Since Pax-5 is known to control the expression of CD19, they do not express that gene, but the block in B-lineage development cannot be attributed to a deficiency in CD19 expression since targeted disruption of CD19 has been shown not to cause B cells developmental arrest at that stage (Rickert et al. 1995). Like preB-I cells from wildtype mice, *Pax-5*$^{-/-}$ preB-I cells can be grown in tissue culture on stromal cells in the presence of the cytokine IL-7 for extended periods of time. They can be cloned in these tissue culture conditions, hence their individual D_HJ_H-rearranged IgH class alleles can be used as genetic markers in all cells that develop from such cloned preB-I cells. *Pax-5*$^{-/-}$ preB-I cells can undergo at least 60 cell divisions without losing their phenotypic and functional properties, certainly more than is needed for genetic modifications and subsequent transplantations.

In order to keep their stem cell properties *Pax-5*$^{-/-}$ preB-I cells have to keep close contact with pre-adipocytic fibroblast-type stromal cells and the cytokine IL-7. Removal of IL-7 from the tissue cultures and concomitant addition of relevant cytokines induces differentiation to different types of hematopoietic cells and reveals some of their pluripotency even *in vitro*. Thus, mere removal of IL-7 without addition of any other cytokine induces these cells to differentiate to macrophages which are capable to phagocytose large particles, even bacteria. This process to macrophage differentiation can be improved by adding the cytokine M-CSF.

Removal of IL-7, followed by addition of G-CSF, allows the development of granulocytes, while addition of GM-CSF, followed by M-CSF, induces differentiation to dendritic cells. These dendritic cells are fully functional antigen-presenting cells, which can stimulate allogeneic T cells to helper and killer T cell responses, and which can take up foreign antigen, process and present it to MHC-restricted, antigen-specific T cells.

Removal of IL-7, followed by addition of IL-2 to the tissue cultures, induces the development of some types of natural killer cells, while culture of the IL-7-

deprived *Pax-5*$^{-/-}$ preB-I cells on stromal cells expressing TRANCE develops osteoclasts which elicit the typical bone-digesting properties. The two latter pathways of differentiation can also be induced *in vivo*. Mice lacking the common γ-chain of the IL-2/IL-7 receptor lack natural killer cells (DiSanto et al. 1995), while c-fos-deficient mice lack osteoclast development (Grigoriadis et al. 1994). Transplantation of *Pax-5*$^{-/-}$ preB-I cells into these deficient mouse strains allows the differentiation of natural killer cells and osteoclasts respectively. This documents that such pluripotent stem cells can fill vacant compartments of the hematopoietic system and, hence, reestablish a fully functional hematopoietic organ with all of its cells.

Finally, transplantation of the *Pax-5*$^{-/-}$ preB-I cells into severe combined immunodeficient *RAG-2*$^{-/-}$ hosts, while not allowing B-cell differentiation, reconstitutes the thymus of the host with all stages of $CD4^-CD8^-$ and $CD4^+CD8^+$ thymocytes, allows normal negative and positive selection of T-cell repertoires on host MHC haplotypes to normally functional $CD4^+$ helper and $CD8^+$ killer T cells in the peripheral lymphoid organs. In the thymus the normal γ/δ TcR T cells develop as well. Again, the *Pax-5*$^{-/-}$ preB-I cells are capable of filling the thymus and the periphery with normal numbers of cells in the previously empty compartments of the T cell lineage.

In all these differentiations to different lineages of the myeloid or T-lymphoid pathways the clones of preB-I cells retain their characteristic D_HJ_H rearrangements on both IgH chain alleles. At the same time they turn off the B-lineage-specific genes, such as VpreB and λ5, and turn on other, lineage-related genes. For instance, in T-cell development they turn on preTα in the thymus, turn it off again, as they begin Vβ to DβJβ rearrangements, followed by Vα to Jα rearrangements, just as in normal T-cell development CD3 genes are expressed.

Remarkably, and again in contrast to preB-I cells from wildtype mice, *Pax-5*$^{-/-}$ preB-I cells are capable of homing back to the bone marrow when transplanted into the immunodeficient hosts. In the transplanted host they retain their state of differentiation, although it is also possible (but not investigated yet) that they dedifferentiate to even earlier stages of hematopoietic cell differentiation. Most excitingly, the donor *Pax-5*$^{-/-}$ preB-I cells in the recipient host can be recloned *ex vivo* on stromal cells in the presence of IL-7 in tissue culture – and display the same pluripotent differentiation capacity *in vitro* and *in vivo* to myeloid and T-lymphoid lineage differentiation. In fact, transplantation of these secondary clones, again, allows them to home to the bone marrow of the secondary host, from where they can be isolated once more as clones of stromal cell/IL-7-responsive, pluripotent cells.

Outlook

The *Pax-5*$^{-/-}$ preB-I-like stem cells can be transgenically influenced since they can be infected with retroviruses *in vitro*. In this way it has been possible to mark the descendants of a *Pax-5*$^{-/-}$ preB cell clone by the expression of green fluorescent protein (GFP), an experimental procedure which allows the subsequent analysis and isolation of marked cells differentiated with *in vitro* or, upon transplantation into a host, *in vivo*. Our ability to retrovirally transfect the *Pax-5*$^{-/-}$ cells should enable the identification of genes by promotor or polyA trap vectors which are active at given stages of development of the different myeloid, natural killer and T-lymphoid lineages.

A variety of genes and their mutants are presently being analysed for their function in T-cell development. Thus, mutants of TcRβ chain genes and of the preTα chain are retrovirally transfected into double mutant *Pax-5*$^{-/-}$/*RAG2*$^{-/-}$, respectively *Pax-5*$^{-/-}$ *pTα*$^{-/-}$ preB-I cells and transplanted into the RAG2$^{-/-}$ hosts for studying the influence of mutated preT-cell receptors on the development of thymocytes at the stages between the DN2 and DN4 types of cells (von Boehmer et al. 1999). A wealth of other combinations of genes and their mutants with suspected or known functions in T-cell development lend themselves to similar studies. It is also conceivable to study genetic requirements for the collaborations of T and B cells in this way by simultaneously transplanting wildtype preB-I cells (infected with suitable genes) yielding B cells only, and *Pax-5*$^{-/-}$ preB-I cells (also suitably genetically modified) yielding T cells only. In this reconstitution of functional lymphoid compartments in the peripheries of the immune system the injection of suitably genetically modified dendritic cells, presenting defined antigens after differentiation *in vitro* from *Pax-5*$^{-/-}$ preB-I cells will hopefully allow assessment of the roles of the initiators in an adaptive immune response of T and B cells.

It is predictable that the analyses of the complexities of the interactions of different cell lineages during the development of the lymphoid cell system, and during responses of T, B and DC cells, will become even more complex as we discover, through DNA chip analysis of genes expressed in given cells at given stages, more and more genes which contribute to this development and these responses.

In summary, *Pax-5*$^{-/-}$ preB-I-like stem cells of the hematopoietic lineages offer the long-awaited opportunity to study the functions of genes in the development and the responses of myeloid, natural killer and T-lymphoid cells *in vitro* and *in vivo*. So far, no abnormalities have been detected in the development of macrophages, granulocytes, osteoclasts, dendritic cells, natural killer cells, thymocytes and mature helper and killer T cells, indicating that the *Pax-5*$^{-/-}$ stem cells should be ideal targets for genetic modifications. Although targeted disruption of genes by homologous recombination has not yet been achieved it is predictable that this technique, used with the *Pax-5*$^{-/-}$ cells, will obviate the need to construct "knock-out" and "knock-in" mice for genes with functions in the various myeloid, natural killer and T-lymphoid cell lineages.

Acknowledgements

The Basel Institute for Immunology was founded and is supported by F. Hoffmann-La Roche Ltd., Basel, Switzerland.

References

DiSanto JP, Muller W, Guy-Grand D, Fischer A, Rajewsky K (1995) Lymphoid development in mice with a targeted deletion of the interleukin 2 receptor gamma chain. Proc Natl Acad Sci USA 92:377-381

Grigoriadis AE, Wang ZQ, Cecchini MG, Hofstetter W, Felix R, Fleisch HA, Wagner EF (1994) c-Fos: a key regulator of osteoclast-macrophage lineage determination and bone remodeling. Science 266:443-448

Morrison SJ, Shah NM, Anderson DJ (1997) Regulatory mechanisms in stem cell biology. Cell 88:287-298

Morrison SJ, Uchida N, Weissman IL (1994) The biology of hematopoietic stem cells. Annu Rev Cell Dev Biol 11:35-71

Morrison SJ, Weissman IL (1994) The long-term repopulating subset of hematopoietic stem cells is deterministic and isolatable by phenotype. Immunity 1:661-673

Nutt SL, Heavey B, Rolink AG, Busslinger M (1999) Commitment to the B-lymphoid lineage depends on the transcription factor Pax5. Nature 401:556-562

Nutt SL, Morrison AM, Dorfler P, Rolink A, M. B (1998) Identification of BSAP (Pax-5) target genes in early B-cell development by loss- and gain-of-function experiments. EMBO J 17:2319-2333

Nutt SL, Urbanek P, Rolink A, Busslinger M (1997) Essential functions of Pax5 (BSAP) in pro-B cell development: difference between fetal and adult B lymphopoiesis and reduced V-to-DJ recombination at the IgH locus. Genes Dev 11:476-491

Rickert RC, Rajewsky K, Roes J (1995) Impairment of T-cell-dependent B-cell responses and B-1 cell development in CD19-deficient mice. Nature 376:352-355

Rodan GA, Harada S (1997) The missing bone. Cell 89:677-680

Rolink AG, Nutt SL, Melchers F, Busslinger M (1999) Long-term in vivo reconstitution of T-cell development by Pax5-deficient B-cell progenitors. Nature 401:603-606

Spangrude GJ, Heimfeld S, Weissman IL (1988) Purification and characterization of mouse hematopoietic stem cells. Science 241:58-62

Urbanek P, Wang ZQ, Fetka I, Wagner EF, Busslinger M (1994) Complete block of early B cell differentiation and altered patterning of the posterior midbrain in mice lacking Pax5/BSAP. Cell 79:901-912

von Boehmer H, Aifantis I, Feinberg J, Lechner O, Saint-Ruf C, Walter U, Buer J, Azogui O (1999) Pleiotropic changes controlled by the pre-T-cell receptor. Curr Opin Immunol 11:135-142

Insertional Mutagenesis as a Route to Identifying Genes Involved in Self Renewal of Haemopoietic Stem Cells

U. Just[1,2], D. Boettiger[1,5], O. Kan[1,6], T. M. Dexter[1,4], and E. Spooncer[1,3]
[1] Paterson Institute for Cancer Research, Wilmslow Road, Manchester M20 9BX, UK
[2] present address: GSF-Institute for Clinical Molecular Biology and Tumorgenetics, Marchioninistr. 25, 81377, Muenchen, Germany
[3] present address: Biomolecular Sciences, UMIST, PO Box 88, Sackville Street, Manchester M60 1QD, UK
[4] present address: The Wellcome Trust, The Wellcome Building, 183 Euston Road, London NW1 2BE, UK
[5] Department of Microbiology, University of Pennsylvania, Philadelphia, Pennsylvania 19104, USA
[6] present address: Oxford Biomedica (UK) Ltd, Medawar Centre, Oxford Science Park, Oxford OX4 4GA, UK

Summary

The genes controlling self renewal in the haemopoietic system are still unknown. Using retroviral insertional mutagenesis we have established multipotent haemopoietic stem cell lines (FDCP-mix) that possess an increased self renewal capacity *in vitro*. To identify genes involved in the regulation of self renewal, proviral integration sites were cloned from FDCP-mix cells and used as probes to screen independently isolated FDCP-mix cell lines for a common proviral insertion site. So far, two common integration sites have been identified, A25 and M4. A25 is rearranged in 50% of the FDCP-mix cell lines and M4 in 10%. Genes located at or near these sites are likely candidates for the control of self renewal of haemopoietic stem cells.

Retroviral Insertional Mutagenesis

Much insight into developmental regulation and oncogenesis has been gained from studies using retroviruses as mutagens (reviewed by Jonkers and Berns 1996; Lock et al. 1991). Retroviruses offer several advantages over conventional mutagenesis induced e.g. by irradiation or chemicals. i) They integrate relatively randomly into the host cell genome at discrete sites as a single integrated provirus; ii) The retrovirus itself provides a molecular tag that facilitates identification and cloning of the affected gene(s); iii) The provirus can activate or inactivate gene(s) located at or nearby the site of integration, thereby creating a mutant phenotype. Multiple mechanisms by which retroviruses can alter the expression of neighbouring genes have been described (Fig. 1). Dependent on the integration site and the transcriptional orientation of the gene, the retroviral regulatory elements located in the 5' and 3' LTR are capable of initiating, enhancing or terminating the transcription of host sequences, resulting in high levels of messenger RNAs encoding the intact protein or the production of aberrant transcripts encoding mutant proteins (Fig. 1). Integration of a retrovirus can also disrupt coding domains or regulatory sequences in such a way that the gene product or its biological activity is lost or reduced.

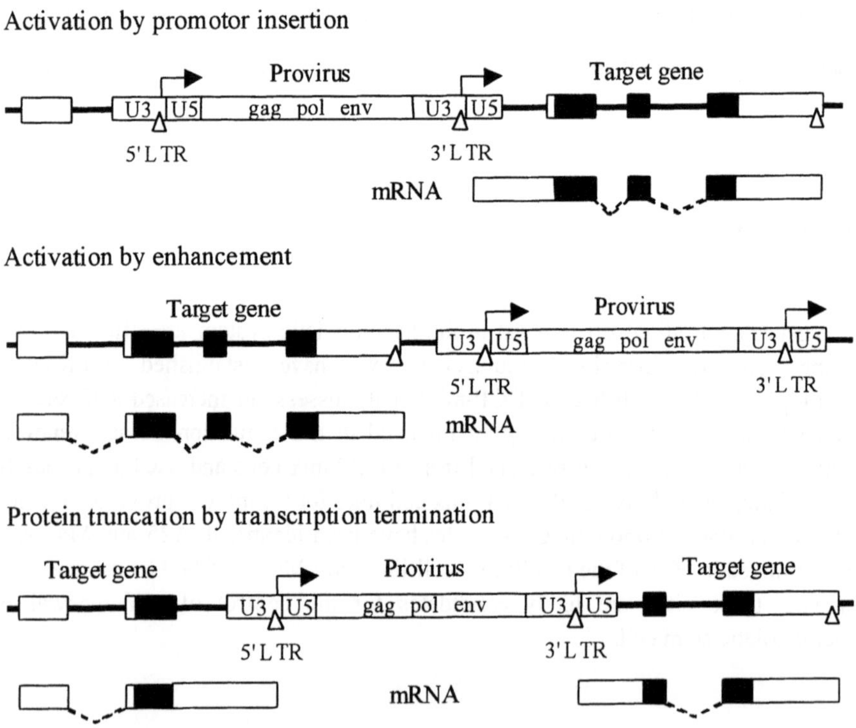

Fig. 1. Mechanisms of retroviral insertional mutagenesis
Structural features of an integrated provirus, and different modes of proviral activation or inactivation of a hypothetical target gene. Protein coding domains are indicated by grey boxes, untranslated regions by open boxes. Arrows define transcriptional start sites and the direction of transcription. Polyadenylation signals are denoted by a triangle. Dotted lines indicate splicing.

We have employed a retroviral insertional mutagenesis strategy to find putative genes that are involved in influencing self renewal of haemopoietic stem cells. Mutants of normal haemopoietic stem cells that show an increased self renewal capacity were isolated from murine long-term bone marrow cultures infected with retroviruses. In these mutant cells we have now identified two retroviral integration sites, A25 and M4, which are rearranged by integration of a retrovirus in 50% and in 10% of the cell lines analysed. These loci are presently being characterised.

FDCP-mix Cells are Clonal Haemopoietic Stem Cell Lines with Increased Self Renewal Capacity

We have established several murine haemopoietic cell lines, named FDCP-mix (factor-dependent cells Paterson mixed potential; Spooncer et al. 1984, 1986), that possess many characteristics of very immature haemopoietic progenitor cells. FDCP-mix cells have a normal karyotype and are non-leukemic. They can be maintained indefinitely *in vitro* either on mouse stromal cells or in the presence of interleukin-3 (IL-3). In the presence of IL-3 they show an immature blast cell morphology (Fig. 2) and express on their cell surface the murine stem cell marker Sca-1 (Ford et al. 1992). FDCP-mix cells have the ability to differentiate in a multilineage response to haemopoietic stromal cells, and to haemopoietic growth factors that promote differentiation of normal primitive haemopoietic cells (Spooncer et al. 1986; Heyworth et al. 1990; Just et al. 1991; Heyworth et al. 1995; T. Schroeder and U. Just unpublished). The range of differentiation

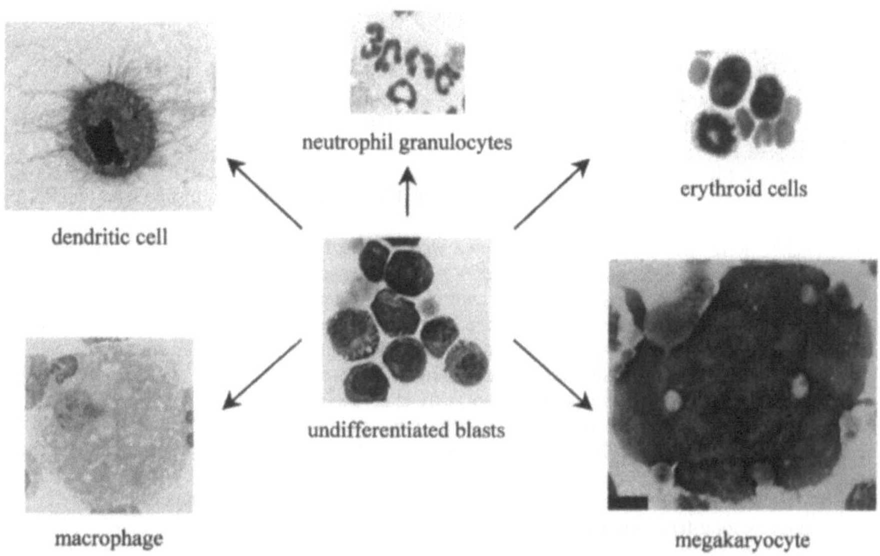

Fig. 2. Multipotential differentiation capacity of FDCP-mix cell lines
FDCP-mix cells were cultured with appropriate cytokines and serum conditions to obtain the cell types shown. Cells were stained with May-Grünwald-Giemsa.

capabilities of these cells *in vitro* includes the granulocyte, macrophage, erythrocyte, megakaryocyte, osteoclast, dendritic cell and B-lymphocyte lineages (Fig. 2; Spooncer et al. 1986; Heyworth et al. 1990; Just et al. 1991; Ford et al. 1992; Heyworth et al. 1995; Hagenaars et al. 1991; T. Schroeder and U. Just unpublished). *In vivo*, FDCP-mix cells form colonies in the spleens of irradiated mice and early isolates can reconstitute haemopoiesis of potentially lethally irradiated mice (Spooncer et al. 1984).

In contrast to normal haemopoietic stem cells, however, FDCP-mix cells have an increased self renewal capacity *in vitro*. As shown in Table 1, normal haemopoietic stem cells gradually lose their self renewal capacity after several passages on normal mouse bone marrow stroma, whereas the balance of self renewal versus differentiation is shifted in favour of self renewal in FDCP-mix cells.

Table 1. Increase in self renewal capacity of src/MCF/MoMuLV infected cultures

		CFU-S transferred[d]	CFU-S recovered[e]	Ratio transferred:recovered
Control[a]				
	Original[c]	180	n.a.	n.a.
	Passage 1	1500	150	10:1
	Passage 2	150	<20	10:1
	Passage 3	<20	0	n.a.
src/MCF/MoMuLV[b]				
	Original[c]	2500	n.a.	n.a.
	Passage 1	1750	1287	1:1
	Passage 2	644	1480	1:2
	Passage 3	740	368	2:1
	Passage 4	184	190	1:1
	Passage 5[f]	95	156	1:2

n.a. not applicable

[a]Long-term bone marrow cultures from adult B6D2F1 mice were established and uninfected cultures were used as controls.

[b]Established long-term bone marrow cultures were infected after 5-8 weeks with a retrovirus mixture consisting of defective src virus, MCF helper virus and MoMuLV.

[c]After 24 weeks, the supernatant cells from control and infected cultures were removed, assayed for CFU-S content and transferred onto irradiated normal mouse bone marrow stroma. The cultures were then maintained for 3 weeks on a weekly demi-depopulation feeding regime. Subsequently, the supernatant cells were collected and transferred onto freshly irradiated mouse bone marrow stroma, and the procedure was repeated at 3-weekly intervals (passages 1-5).

[d]Number of CFU-S from established long-term bone marrow cultures or passages (uninfected control cultures or src/MCF/MoMuLV infected cultures) transferred onto irradiated mouse bone marrow stroma.

[e]Upon each transfer, the supernatants were assayed for CFU-S (total recovered per flask).

[f]Cells possess long-term repopulation capacity.

FDCP-mix Cells Arise as a Consequence of Infection with Replication Competent Helper Virus

The FDCP-mix cell lines were established from long-term bone marrow cultures infected with a defective vector expressing the src-oncogene, and MoMuLV and/or MCF helper virus (Table 2; Boettiger et al. 1984; Spooncer et al. 1984). The infection of the marrow cultures with these viruses resulted in an increased production of haemopoietic progenitor cells. However, the FDCP-mix cells themselves do not harbour the src-virus (Wyke et al. 1986). Thus, the influence of the src oncogene on establishment of the FDCP-mix cells is only indirect. Most likely, immature haemopoietic progenitor cells were stimulated to proliferate by the stromal cells expressing the src oncogene. Furthermore, cell lines with a similar phenotype to FDCP-mix cells were obtained from haemopoietic tissues of mice infected with MoMuLV and a defective vector expressing IL-3 (Just et al. 1993). Thus, the high proliferative activity in these systems then may allow the helper virus to act as an insertional mutagen.

Table 2. Generation of clonal multipotent progenitor cell lines (FDCP-mix) is a consistent but rare event

	Retrovirus used with the 2.1 src virus[a]		
Experiment[b]	MCF + MoMuLV	MCF	MoMuLV
A1-A10	>50	n.d.	n.d.
1	0	5	0
2	14	2	0
3	11/21[c]	n.d.	n.d.
4	3	n.d.	n.d.
5	1	2	n.d.
6	0	4	2
total	>50	13	2

n.d. not done

[a]Long-term bone marrow cultures from adult mice were established and infected after 5-8 weeks with a retrovirus mixture consisting of the defective 2.1 src virus and MCF helper virus and/or MoMuLV. Shown is the number of clonal cell lines generated.

[b]Long-term marrow cultures were established from B6D2F1 mice, except for experiment 6 in which bone marrow from $HGPRT^{null}$ mice was used.

[c]11 individual cultures out of a total of 21 infected cultures yielded a FDCP-mix cell line. The haemopoietic cells in the cultures were already clonal at the time of cell line establishment. Considering the number of progenitor cells produced during culture time, the estimated frequency of obtaining a FDCP-mix type cell line is less than 10^{-6}.

To determine the role of the helper virus in the establishment of the FDCP-mix cell lines, we analysed more than 20 independently isolated FDCP-mix clones for their proviral integration pattern using MoMuLV or MCF virus-specific probes (Stoye and Coffin 1987). Between 4 to 7 proviral inserts were found per cell (Fig. 3). Most clonally derived cell lines out of an experiment had the same integration pattern. In addition, establishment of FDCP-mix cells was a rare event. This indicates that insertion of the provirus in specific sites of host DNA could be causally related to the ability of the FDCP-mix cell lines to self renew.

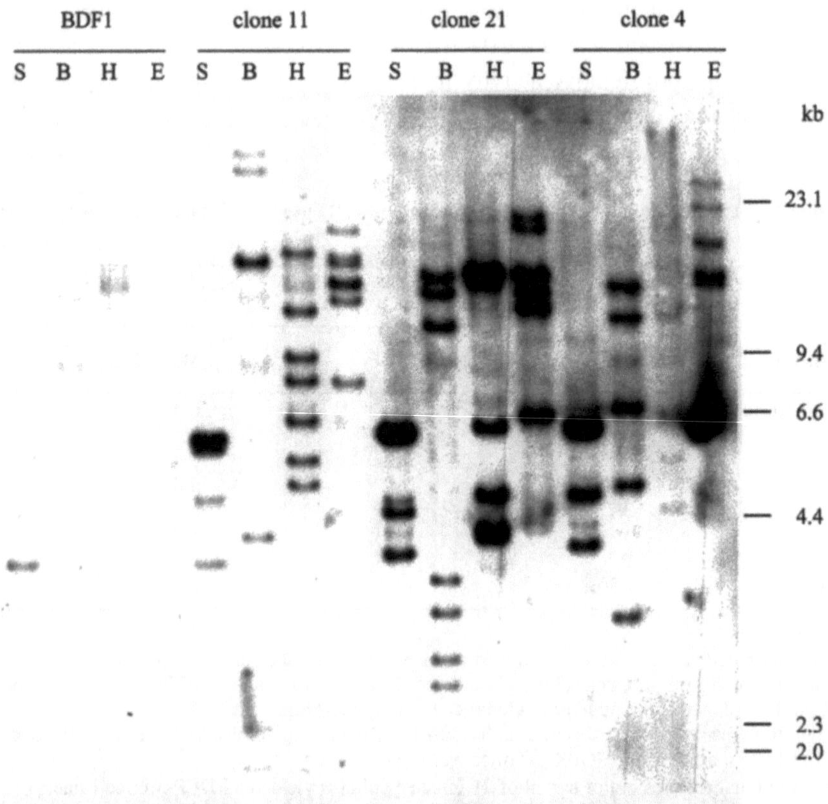

Fig. 3. Retroviral integration sites in FDCP-mix cell lines
Southern blot analysis of genomic DNA of FDCP-mix cells. DNA was digested with EcoRI (E), which does not cut within the viral genome, with HindIII (H) and BamHI (B), which cut once within the viral genome, and with SstI (S) which cuts several times within the MoMuLV genome. Hybridisation with a MoMuLV env specific probe yields a single fragment with EcoRI, HindIII and BamHI for each integrated MoMuLV, and a 5.7 kb fragment for the MoMuLV used for infection. Between 4 to 7 integration MoMuLV sites were detected in FDCP-mix cell lines. Shown are integrations for FDCP-mix clones #11, #21, and #4. Spleen DNA from mice of the relevant genetic background (BDF1) was used as control.

A25 and M4 are Common Retroviral Integration Sites in FDCP-mix Cell Lines

In the search for a common retroviral integration site, i.e., a genomic locus at which a retrovirus is integrated in several independently derived FDCP-mix cell lines, we cloned all retroviral integration sites out of one cell line, FDCP-mix clone #7. Genomic DNA was fractionated and the four fractions each containing one of the proviral integration sites were used in an one-sided PCR strategy to amplify the flanking sequences. The cloned flanking sequences were then used as probes on Southern blots to screen for rearrangements in FDCP-mix cell lines. As shown in Fig. 4, one of the four integration sites of FDCP-mix clone #7 was rearranged in several independently isolated FDCP-mix cell lines. In total, about 50% of the FDCP-mix cell lines (Table 3) showed a rearrangement caused by a retrovirus (either MoMuLV or MCF virus) in this region (A25). Since not all FDCP-mix cell lines were rearranged at the A25 region, we reasoned that other common integration sites may exist. We therefore cloned several retroviral integration sites of another FDCP-mix cell line, clone #4. One of these integration sites (M4) was commonly rearranged in about 10% of FDCP-mix cell lines (Table 3). Further work is in progress to search for additional common integration sites.

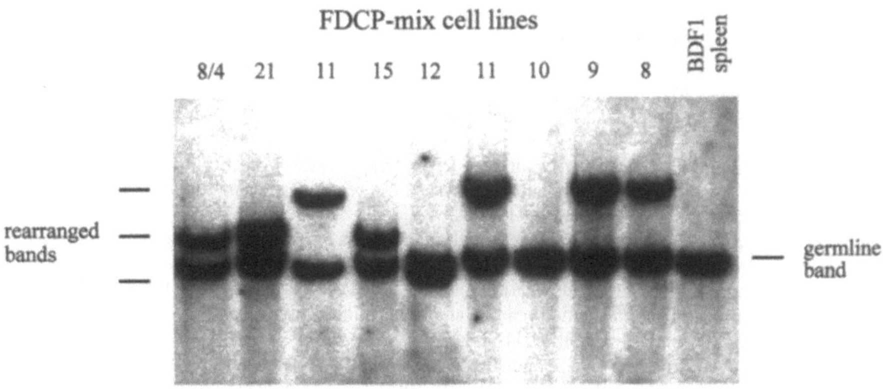

Fig. 4. Rearrangements within the A25 locus
Southern blot of HindIII digests of the indicated FDCP-mix cell line DNA or of spleen DNA from mice with the relevant background (BDF1) were hybridised with an A25 probe.

Table 3. Common integration sites in FDCP-mix cell lines

Locus	Cell lines analysed	Cell lines rearranged	Percent rearranged
A25	31	17	55
M4	31	3	10

The high incidence of retroviral integration in FDCP-mix cell lines into the A25 or M4 sites indicates that genes located at or nearby to these areas are likely to be involved in generating the FDCP-mix phenotype, which is an increased self renewal capacity *in vitro*. We are currently cloning the genes located in these regions. Further analysis will show if and which of these gene(s) are involved in the regulation of self renewal of haemopoietic stem cells.

Acknowledgements
This work was supported by the Deutsche Forschungsgemeinschaft, Germany (Ju 197/3-1, Ju 197/3-2 and Ju 197/4-1 to UJ) and the Leukemia Research Fund, UK (to ES). ES is supported by the Cancer Research Campaign, UK. During the course of this work, DB was on sabbatical leave in Manchester from the University of Pennsylvania. The authors thank Mrs J. Fenelly, Mrs D. Gagen and Mrs G. Johnson for expert technical assistance.

References

Boettiger D, Anderson S, Dexter M (1984) Effect of src infection on long-term marrow cultures: increased self-renewal of hemopoietic progenitor cells without leukemia. Cell 36:763-773

Ford A, Healy L, Bennett A, Navarro E, Spooncer E, Greaves M (1992) Multilineage phenotype of interleukin-3-dependent progenitor cells. Blood 79:1962-1971

Hagenaars C, Kawilarang de Haas E, van der Kraan A, Spooncer E, Dexter M, Nijweide P (1991) Interleukin-3-dependent hematopoietic stem cell lines capable of osteoclast formation in vitro. J. Bone Miner. Res. 6:947-954

Heyworth C, Dexter M, Kan O, Whetton A (1990) The role of hemopoietic growth factors in self-renewal and differentiation of IL-3-dependent multipotential stem cells. Growth Factors 2:297-211

Heyworth C, Alauldin M, Cross M, Fairbairn L, Dexter M (1995) Erythroid development of the FDCP-mix A4 multipotent cell line is governed by relative concentrations of erythropoietin and interleukin 3. Br. J. Haematol. 91:15-22

Jonkers J, Berns A (1996) Retroviral insertional mutagenesis as a strategy to identify cancer genes. Biochim. Biophys. Acta 1287:29-57

Just U, Stocking C, Spooncer E, Dexter M, Ostertag W (1991) Expression of the GM-CSF gene after retroviral transfer in hematopoietic stem cell lines induces synchronous granulocyte-macrophage differentiation. Cell 64:1163-1173

Just U, Katsuno M, Stocking C, Spooncer E, Dexter M (1993) Targeted in vivo infection with a retroviral vector carrying the Interleukin-3 (Multi-CSF) gene leads to immortalization and leukemic transformation of primitive hematopoietic progenitor cells. Growth Factors 9:41-55

Lock L, Jenkins N, Copeland N (1991) Mutagenesis of the mouse germline using retroviruses. Curr. Top. Microbiol. Immunol. 171:27-41

Spooncer E, Boettiger D, Dexter M (1984) Continuous *in vitro* generation of multipotent stem cell clones from src-infected cultures. Nature 310:228-230

Spooncer E, Heyworth CM, Dunn A, Dexter M (1986) Self-renewal and differentiation of interleukin-3-dependent multipotent stem cells are modulated by stromal cells and serum factors. Differentiation 31:111-118

Stoye J, Coffin J (1987) The four classes of endogenous murine leukemia virus: structural relationships and potential for recombination. J. Virol. 61:2659-2669

Wyke J, Stoker A, Searle S, Spooncer E, Simmons P, Dexter M (1986) Perturbed hemopoiesis and the generation of multipotential cell clones in src-infected bone marrow cultures is an indirect or transient effect of the oncogene. Mol. Cell. Biol. 6:959-963

The Helix-Loop-Helix Inhibitor Id2 and Cell Differentiation Control

Y. Yokota[1], S. Mori[1], S.-I. Nishikawa[1], A. Mansouri[2], P. Gruss[2], T. Kusunoki[3], T. Katakai[4] and A. Shimizu[4]
[1] Department of Molecular Genetics, Kyoto University Graduate School of Medicine, 606-8507 Kyoto, Japan
[2] Department of Molecular Cell Biology, Max-Planck Institute of Biophysical Chemistry, Am Fassberg, D-37077 Göttingen, Germany
[3] Department of Pediatrics, Kyoto University Graduate School of Medicine, 606-8507 Kyoto, Japan
[4] Center for Molecular Biology and Genetics, Kyoto University, 606-8507 Kyoto, Japan

HLH Factors

Transcription factors play pivotal roles not only in the regulation of cellular functions but also in the control of cell differentiation. They can be classified on the basis of their structural features, i.e. motifs that mediate functional properties such as DNA binding and protein-protein interactions. The family of helix-loop-helix (HLH) factors is one of the well-studied examples [1-3]. Most members have DNA binding activity and directly regulate the expression of their downstream genes. Some, however, do not bind DNA and negatively regulate the functions of other HLH transcription factors at the protein level [4, 5]. These HLH inhibitors are called Id, inhibitor of DNA binding/differentiation [4, 5]. Here we focus on and describe the immunological aspects of mice lacking Id2.

bHLH Transcription Factors and HLH Inhibitors

Members of the expanding family of basic helix-loop-helix (bHLH) transcription factors are characterized by two conserved regions; one rich in basic amino acids and the other containing the helix-loop-helix [1-3]. The former is required for DNA binding and the latter for the formation of homo- or heterodimers. The dimerization has been shown to be critical for the transcriptional activation function [1-3]. Among the best known examples are the myogenic bHLH factors, MyoD and myogenin, that are involved in skeletal muscle development [1, 6], SCL/TAL1 in hematopoiesis [7] and Mash1 and neurogenin in neurogenesis [8, 9]. These cell differentiation processes are regulated not only by bHLH factors but also by negative regulators of bHLH factors, the Id proteins [4, 5].

Id proteins possess the characteristic HLH motif and can heterodimerize with bHLH factors, particularly with the ubiquitously expressed class A molecules such as E2A gene products, thereby exerting a "quenching out" effect on the dimerization partners of tissue-specific bHLH factors [4, 5]. In addition, the resultant heterodimers cannot bind DNA because of the lack of the basic region in Id [4, 5]. Id proteins thus negatively regulate the functions of bHLH factors and cell differentiation. Moreover, Id proteins have been demonstrated to stimulate cell cycle progression [5]. These properties implicate a role for Id proteins in the expansion of undifferentiated cell populations during the process of development [5], but detailed in vivo functions are still unclear.

Mice Deficient for Id2

In contrast to the large number of bHLH factors that have been identified, to date only 4 types of Id proteins have been reported in mammals. This suggests that each member of the Id family regulates the functions of multiple bHLH molecules in the process of cell differentiation. To investigate the in vivo functions of Id proteins, we focused on Id2 [10] and generated mice deficient for this factor through conventional homologous recombination in embryonic stem cells. Id2-deficient mice exhibit several immunological defects as discussed below [11], in addition to other phenotypes including a lactation defect [Mori et al., submitted].

Alymphoplasia in Id2-Deficient Mice

Peripheral lymphoid organs are sites for immune reactions and consist of lymph nodes, Peyer's patches and spleen. Examination of Id2-deficient mice revealed a lack of lymph nodes and Peyer's patches [11]. Transplantation of $Id2^{-/-}$ bone marrow cells into lethally irradiated wild type mice demonstrated the normal homing activity of $Id2^{-/-}$ lymphocytes to lymph nodes and Peyer's patches, while transfer of wild type bone marrow cells into Id2-deficient mice did not induce the formation of lymph nodes or Peyer's patches [11]. These results suggest that the places for lymphocyte homing are missing in Id2-deficient mice.

Immunohistochemical studies have shown that developing Peyer's patches can be identified as VCAM-1 positive spots from embryonic day 15 onward; VCAM-1 stands as the earliest marker for developing Peyer's patches [12]. However, VCAM-1 positive spots could not be detected in the intestines of Id2-deficient mice [11], indicating that Peyer's patch development is disturbed at very early stages.

It has been shown that signal transduction through the lymphotoxin (LT) receptor β is critical for the development of peripheral lymphoid organs [13 - 18]. Mice lacking LTα or β exhibit defects in the formation of peripheral lymphoid organs [13, 15]. Inactivation of the LTβ receptor by either gene targeting or administration of a chimeric LTβ receptor-IgG protein also results in a similar phenotype [14, 16, 17]. We therefore investigated the localization patterns of LT mRNAs in neonatal digestive tracts. In control neonates, transcripts of both LTα and LTβ were detected as spots on the antimesenteric side of the intestines, similar to spots positive for VCAM-1 expression visualized by immunohistochemistry, indicating the localization of LT transcripts to Peyer's patch cells [11]. In Id2-deficient mice, no LT-expressing cells could be detected [11].

For the LT-expressing cells, Mebius et al. have reported a specific cell population which is positive for CD4 and IL-7 receptor-α (IL-7Rα), and negative for CD3 [19]. This population is found in embryonic peripheral lymphoid organs, peripheral blood [19] and also in embryonic intestines [20]. In addition to the expression of LT, the spatial and temporal emergence of this cell population during development suggests an essential role in the generation of peripheral lymphoid organs [19, 20]. We investigated whether this $CD4^+CD3^-IL\text{-}7R\alpha^+$ cell population is present in Id2-deficient mice. Flow cytometric analyses identified this cell population in the dispersed intestinal cells of control littermates at an abundance of approximately 2.0% but these cells were not detectable in the Id2-deficient mice [11], suggesting that lack of this specific cell population may underlie the alymphoplasia phenotype of the mutant mice. Reciprocally, Id2 null mutant mice highlight the important role of this cell population in the formation of lymph nodes and Peyer's patches.

Normal Splenic Structure and Function in Id2-Deficient Mice

Defective formation of lymph nodes and Peyer's patches is usually accompanied by disorganized splenic architecture and impaired humoral responses as seen in mice deficient for LTα, LTβ or LTβ receptor [18]. The impairment in immune function is thought to be a reflection of the abnormal splenic structure of these mutant mice. In Id2 mutant spleens, however, clear segregation of T and B cells and distinct formation of germinal centers are observed [11]. Furthermore, marginal metallophilic macrophages and follicular dendritic cells can be found in the mutants, similar to control spleens [11 and Yokota, unpublished observation]. Id2-deficient mice show normal serum levels of IgM, IgG and IgA, and exhibit sufficient induction of T cell-dependent B cell responses against sheep red blood cells and a hapten [11]. Thus, Id2-deficient mice have an apparently normal structure and function of the spleen, despite the alymphoplasia phenotype. These results indicate that the spleen is different in its dependence on the LT-expressing cells from lymph nodes and Peyer's patches.

Impaired NK Cell Differentiation in Id2-Deficient Mice

In addition to the defects in peripheral lymphoid organ development, Id2-deficient mice have greatly reduced populations of natural killer (NK) cells in the spleen, liver and bone marrow, 10-20% compared to those of control littermates [11]. $Id2^{-/-}$ splenocytes exhibit a cytolytic activity comparable to the level of NK cells present [11]. It is known that NK cell development depends on both NK cell precursors and the microenvironment in the bone marrow. Previous studies have demonstrated that the microenvironment of the bone marrow can be mimicked by the addition of IL-15 into culture media [21, 22]. After 7 days of culture with IL-15, 25% of $Id2^{+/-}$ bone marrow cells were NK cells, whereas less than 1% were so in $Id2^{-/-}$ bone marrow cells [11]. This result suggests that Id2-deficient mice also have a defect in NK cell progenitors. For confirmation, bone marrow transplantation was performed: $Id2^{+/-}$ and $Id2^{-/-}$ bone marrow cells that were transferred into lethally irradiated wild type mice generated NK cells at 1.6% and 0.4% of splenocytes respectively [11]. After transplantation of wild type bone marrow cells into $Id2^{+/-}$ and $Id2^{-/-}$ recipient mice, on the contrary, NK cells were detected at 2.6% and 1.7% of splenocytes, respectively [11]. These experiments demonstrated that a defect in the NK cell progenitors is likely responsible for the impaired differentiation of NK cells in Id2-deficient mice.

Other Findings

Flow cytometric analyses further identified increased and decreased percentages of CD4 and CD8 single positive T cells, respectively [Yokota, unpublished observation]. In the $Id2^{-/-}$ spleen, CD4 and CD8 single positive T cells are respectively double and half of those in control spleens, resulting in an increased CD4/CD8 ratio of more than 12.

As described above, Id2-deficient mice have normal levels of serum IgM, total IgG and IgA. Detailed analyses, however, identified a greatly elevated serum IgE level: a 20-fold increase compared to that of control littermates [Kusunoki et al., submitted]. This elevated IgE level is accompanied by an increase of IgG1 and decrease of IgG2b levels, indicating that the Id2 KO mice are in a Th2-dominant state [Kusunoki et al., submitted]. Moreover, isolated $CD4^+$ naive T cells show impaired differentiation to Th1 cells, even in conditions that preferentially support Th1 cell differentiation [Kusunoki et al., submitted]. However, we cannot exclude the possibility of B cell involvement, since the impaired differentiation ability of $CD4^+$ naive T cells does not seem severe enough to fully explain the elevated serum IgE level. The effect of E-proteins, dimerization partners of Id, on class switching in B cells has been reported [23].

Role of Id2 in the Murine Immune System

As described here, Id2-deficient mice show a variety of defects in the immune system. How do these phenotypes arise? Id proteins are negative regulators of cell differentiation and stimulators of cell proliferation; the former is considered to be the major Id function [5]. Therefore, at the moment, it is natural to speculate the presence of a bHLH factor at the diverging point of cell differentiation.

$CD4^+CD3^-IL\text{-}7R\alpha^+$ Cells and NK Cells

One may speculate that Id2 supports the generation of LT-expressing $CD4^+CD3^-IL\text{-}7R\alpha^+$ cells from precursors in the lymphoid lineage through suppression of the activity of a yet unknown bHLH factor(s) that promotes differentiation to other lineage(s) as illustrated in Fig. 1. LT-expressing $CD4^+CD3^-IL\text{-}7R\alpha^+$ cells have been shown to differentiate to NK cells under appropriate in vitro culture conditions [19]. Thus the defective differentiation of LT-expressing $CD4^+CD3^-IL\text{-}7R\alpha^+$ cells might explain both the defective development of peripheral lymphoid organs and impaired NK cell differentiation (Model 1 in Fig. 1). Alternatively, it is also possible that Id2 is independently essential for $CD4^+CD3^-IL\text{-}7R\alpha^+$ cells and NK cells (Model 2 in Fig. 1). LT-expressing $CD4^+CD3^-IL\text{-}7R\alpha^+$ cells are detected only during the embryonic and neonatal periods [19, 20], while transplantation of adult bone marrow cells can reconstitute NK cells in recipients. Moreover, bipotent T/NK cell progenitors are found in the fetal thymus [24] where $CD4^+CD3^-IL\text{-}7R\alpha^+$ cells are not detected [19]. Based on this finding, we tend to prefer model 2.

Regarding NK cell development, it has been reported that overexpression of Id3 in human $CD34^+$ fetal thymocytes promotes the generation of NK cells at the expense of T cells in the fetal thymus organ culture system [25]. This provides supporting evidence for a role of Id proteins in the differentiation of NK cells from bipotential T/NK progenitors. It is highly probable that a bHLH factor facilitates commitment of bipotent T/NK progenitors to the T cell lineage and functional inactivation of that factor leads to differentiation of NK cells.

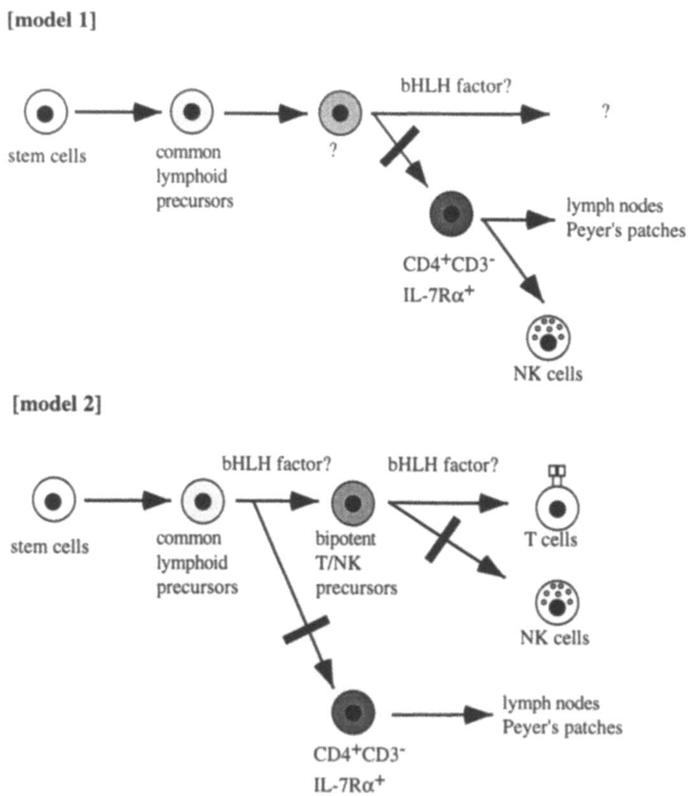

Fig. 1. Schematic models of Id2 in development of peripheral lymphoid organs and NK cells. Model 1 proposes that the defects in peripheral lymphoid organs and NK cells are caused by a single origin. In model 2, two independent impaired steps are proposed to generate the respective defects. In both cases, involvement of a bHLH factor can be speculated at each diverging point.

T Cell Development

The mechanisms underlying the defects in the $Id2^{-/-}$ T cell population are still unclear. In respect to the increased ratio of CD4/CD8, we should consider both T cells and environment, although it is straightforward and easy to speculate that factors intrinsic to T cells are the cause. Preliminary cell transfer experiments have demonstrated that $Id2^{-/-}$ bone marrow cells do not reproduce the deviation in the CD4/CD8 ratio [Yokota, unpublished observation], suggesting that the thymic environment may be responsible for the deviation in Id2-deficient mice. There is no evidence at present showing the expression of Id2 in the thymic stromal cells, although Northern blot analyses indicate abundant expression of Id2 in the thymus [Yokota, unpublished observation].

Differentiation of $CD4^+$ naive T cells to Th1 and Th2 cells involves many players including cytokines such as INF-γ and IL-4 [26]. RT-PCR analyses indicate that cultured Th1 and Th2 cells express Id2 at a similar level [Yokota et al., unpublished observation]. The fact that $CD4^+$ naive T cells show impaired ability to differentiate into Th1 cells implies that Id2 may participate in proliferation and/or survival of Th1 cells, as shown in other cell types [5].

Conclusion

Genetically engineered mice sometimes provide us with unexpected clues to understand mechanisms underlying complex biological processes and phenomena. Mice lacking Id2 have revealed its importance in several processes that generate important components in the immune system. Further issues to be resolved would certainly include factors that are regulated by Id2. Identification of these molecules will help our understanding of how cell differentiation is controlled in the immune system. In addition, we may need to focus more attention to the role of Id in cell proliferation and survival in the immune system.

Acknowledgment

We thank Ruth T. Yu for reviewing the manuscript. This work was supported by the Max-Planck Society, Germany, the Ministry of Education, Science, Sports and Culture of Japan and the Yamanouchi Foundation.

References

1. Weintraub, H., Davis, R., Tapscott, S., Thayer, M., Krause, M., Benezra, R., Blackwell, T. K., Turner, D., Rupp, R., Hollenberg, S., Zhuang, Y. and Lassar, A. (1991). The myoD gene family: nodal point during specification of the muscle cell lineage. Science 251: 761-766.
2. Massari, M. E. and Murre, C. (2000). Helix-loop-helix proteins: regulators of transcription in eucaryotic organisms. Mol. Cell. Biol.. 20: 429-440.
3. Littlewood, T. and Evan, G. (1998). Helix-Loop-Helix Transcription Factors. 3rd Edition (Oxford, Oxford University Press).
4. Benezra, R., Davis, R. L., Lockshon, D., Turner, D. L. and Weintraub, H. (1990). The protein Id: a negative regulator of helix-loop-helix DNA binding proteins. Cell 61: 49-59.
5. Norton, J. D., Deed, R. W., Craggs, G. and Sablitzky, F. (1998). Id helix-loop-helix proteins in cell growth and differentiation. Trends Cell Biol. 8: 58-65.
6. Molkentin, J. D. and Olson, E. N. (1996). Defining the regulatory networks for muscle development. Curr. Opin. Genet. Dev. 6: 445-53.
7. Jordan, C., and Van Zant, G. (1998). Recent progress in identifying genes regulating hematopoietic stem cell function and fate. Curr. Op. Cell Biol. 10, 716-720.
8. Lee, J. E. (1997). Basic helix-loop-helix genes in neural development. Curr. Opin. Neurobiol. 7: 13-20.

9. Kageyama, R. and Nakanishi, S. (1997). Helix-loop-helix factors in growth and differentiation of the vertebrate nervous system. Curr. Opin. Genet. Dev. 7: 659-665.
10. Sun, X. H., Copeland, N. G., Jenkins, N. A. and Baltimore, D. (1991). Id proteins Id1 and Id2 selectively inhibit DNA binding by one class of helix-loop-helix proteins. Mol. Cell. Biol. 11: 5603-5611.
11. Yokota, Y., Mansouri, A., Mori, S., Sugawara, S., Adachi, S., Nishikawa, S.-I. and Gruss, P. (1999). Development of peripheral lymphoid organs and natural killer cells depends on the helix-loop-helix inhibitor Id2. Nature 397: 702-706.
12. Adachi, S., Yoshida, H., Kataoka, H. and Nishikawa, S.-I. (1997). Three distinctive steps in Peyer's patch formation of murine embryo. Int. Immunol. 9: 507-514.
13. De Togni, P., Goellner, J., Ruddle, N. H., Streeter, P. R., Fick, A., Mariathasan, S., Smith, S. C., Carlson, R., Shornick, L. P., Strauss-Schoenberger, J., Russell, J. H., Karr, R. and Chaplin, D. D. (1994). Abnormal development of peripheral lymphoid organs in mice deficient in lymphotoxin. Science 264: 703-707.
14. Rennert, P., Browning, J., Mebius, R., Mackay, F. and Hochman, P. (1996). Surface lymphotoxin α/β complex is required for the development of peripheral lymphoid organs. J. Exp. Med. 184: 1999-2006.
15. Koni, P. A., Sacca, R., Lawton, P., Browning, J. L., Ruddle, N. H. and Flavell, R. A. (1997). Distinct roles in lymphoid organogenesis for lymphotoxins α and β revealed in lymphotoxin β -deficient mice. Immunity 6: 491-500.
16. Fütterer, A., Mink, K., Luz, A., Kosco-Vilbois, M. H. and Pfeffer, K. (1998). The lymphotoxin β receptor controls organogenesis and affinity maturation in peripheral lymphoid tissues. Immunity 9: 59-70.
17. Rennert, P. D., James, D., Mackay, F., Browning, J. L. and Hochman, P. S. (1998). Lymph node genesis is induced by signaling through the lymphotoxin β receptor. Immunity 9: 71-79.
18. Chaplin, D. and Fu, Y. (1998). Cytokine regulation of secondary lymphoid organ development. Curr. Op. Immunol. 10: 289-97.
19. Mebius, R., Rennert, P. and Weissman, I. (1997). Developing lymph nodes collect $CD4^+CD3^-LT\beta^+$ cells that can differentiate to APC, NK cells, and follicular cells but not T or B cells. Immunity 7: 493-504.
20. Yoshida, H., Honda, K., Shinkura, R., Adachi, S., Nishikawa, S., Maki, K., Ikuta, K. and Nishikawa, S.-I.. (1999). IL-7 receptor α^+ $CD3^-$ cells in the embryonic intestine induces the organizing center of Peyer's patches. Int. immunol. 11: 643-655.
21. Carson, W. E., Giri, J. G., Lindemann, M. J., Linett, M. L., Ahdieh, M., Paxton, R., Anderson, D., Eisenmann, J., Grabstein, K. and Caligiuri, M. A. (1994). Interleukin (IL) 15 is a novel cytokine that activates human natural killer cells via components of the IL-2 receptor. J. Exp. Med. 180: 1395-1403.
22. Ogasawara, K., Hida, S., Azimi, N., Tagaya, Y., Sato, T., Yokochi Fukuda, T., Waldmann, T. A., Taniguchi, T. and Taki, S. (1998). Requirement for IRF-1 in the microenvironment supporting development of natural killer cells. Nature 391: 700-703.
23. Quong, M., Harris, D., Swain, S. and Murre, C. (1999). E2A activity is induced during B-cell activation to promote immunoglobulin class switch recombination. EMBO J. 18: 6307-6318.
24. Rodewald, H., Moingeon, P., Lucich, J., Dosiou, C., Lopez, P. and Reinherz, E. (1992). A population of early fetal thymocytes expressing Fc γ RII/III contains precursors of T lymphocytes and natural killer cells. Cell 69: 139-150.
25. Heemskerk, M. H., Blom, B., Nolan, G., Stegmann, A. P., Bakker, A. Q., Weijer, K., Res, P. C. and Spits, H. (1997). Inhibition of T cell and promotion of natural killer cell development by the dominant negative helix-loop-helix factor Id3. J. Exp. Med. 186: 1597-1602.
26. Coffman, R., Mocci, S. and O'Garra, A. (1999). The stability and reversibility of Th1 and Th2 populations. Curr. Top. Microbiol. Immunol. 238: 1-12.

Role of β1 Integrin for Hemato-Lymphopoiesis in Mouse Development

A. J. Potocnik

Basel Institute for Immunology, Grenzacherstrasse 487, CH-4005 Basel, Switzerland

Introduction

Integrins are heterodimeric cell adhesion molecules with major roles in a variety of biological processes ranging from cell migration to tissue organization, growth, differentiation, and programmed cell death. Their main biological function is defined by mediating cell-cell and cell-extracellular matrix contacts and transmitting signals in both directions. Each integrin is composed of an α and a β subunit assembling noncovalently in a variety of combinations. To date, 16 α and 8 β subunits are known forming at least 22 different integrins (Hynes, 1992). Subfamilies of integrins are named according to the β chain utilized. The largest family comprises the β1 integrins consisting of at least twelve members. The complexity of the β1 integrin family is further increased by the presence of alternative splicing of mRNAs which occurs in genes encoding α as well as β subunits (van der Flier et al., 1995; Zhidkova et al., 1995; Belkin et al., 1996). Heterodimers of the β1 integrin subfamily bind various components of the extracellular matrix (ECM) including collagens (α1, α2), laminins (α1, α2, α3, α6, α7), fibronectin (α3, α4, α5, α8, αv), tenascin (α9), vitronectin (αv) as well as cell counter receptors such as vascular cell adhesion molecule 1 (VCAM-1) (α4; see Fig. 1).

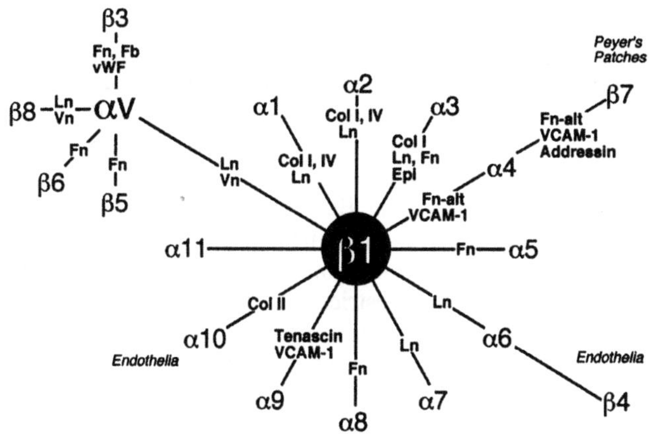

Fig. 1. Integrin subunits and ligands. The known pairings of α and β subunits to form the final heterodimer are depicted by the lines, on which are listed the specific ligand bound by that heterodimer. Col = collagen; Fb = fibrinogen; Fn = fibronectin; Fn-alt = fibronectin alternatively spliced region; Ln = laminin; Vn = vitronectin; vWF = von Willebrand factor.

Despite the promiscuity of ligand-receptor combination each integrin can transduce unique information (Wary et al., 1996) by alteration of the affinity for the respective extracellular ligand. Affinity changes are dependent upon cellular activation mechanisms eventually resulting in a conformational change of the integrin - a process termed "inside-out signaling" (Schwartz et al., 1995; Dedhar and Hannigan, 1996). In addition, integrins can trigger several signaling mechanisms including the mitogen-activated protein kinase cascade (Chen et al., 1994; Schlaepfer et al., 1994), protein kinase C (Vuori and Ruoslahti, 1993), phosphatidylinositol hydrolysis (McNamee et al., 1993), Na^+/H^+ antiporter activity (Schwartz et al., 1991) and, as a very early event, a pathway involving Rho (Chong et al., 1994).

A large set of functions both in the developing as well as in the adult organism has been assigned to integrins. Specifically targeted mutations of integrin genes in the mouse have contributed to a better understanding of their physiological roles (Fässler et al. 1996; Hynes, 1996). So, these families of cell adhesion molecules play crucial roles in hematopoiesis, vasculogenesis, angiogenesis and organogenesis. In this review we will focus on the function of β1 integrins for the development, differentiation and migration of hematopoietic cells during embryogenesis.

Function of β1 Integrin on Hematopoietic Progenitors during Embryogenesis

Generation of the Primitive and Definitive Hematopoietic System

The first visible sign of hematopoietic activity in the mouse embryo is the appearance of blood islands in the developing yolk sac at approximately E7.5 (Russel 1979). Within these yolk sac blood islands, a population of primitive, nucleated erythroid cells is surrounded by a layer of mesodermal cells that eventually give rise to angioblasts forming the developing vasculature. The parallel development of these lineages in close association in the blood islands of the yolk sac provided the basis for the hypothesis that they arise from a common precursor, the so-called hemangioblast (Sabin 1920). In addition to this extra-embryonic site of primitive hematopoiesis, hematopoietic progenitors are generated at an intra-embryonic site: the para-aortic splanchnopleura (p-Sp)/aorta-gonad-mesonephros (AGM) region. Before the amalgamation of the vascular system and the initiation of heart beating at approximately E8.5 (8-9 somite pairs), both the yolk sac and the p-Sp contain myeloid and erythroid progenitors whereas the lymphoid potential is restricted to the intra-embryonic p-Sp (Cumano et al., 1996). Accumulating evidence suggests that the potential for definitive hematopoiesis is primarily restricted to the p-Sp (Medvinsky et al., 1993; Godin et al., 1995; Medvinsky and Dzierzak, 1996), whereas the yolk sac represents the main site of primitive pre-fetal liver hematopoiesis (Keller et al., 1999). The fetal liver rudiment is colonised by hematopoietic stem cells (HSCs) starting at E9.5-10 (Houssaint 1981). Until birth the fetal liver remains the main site of definitive hematopoiesis in the embryo. Analogous to the fetal liver, also fetal thymus

(E10.5), spleen (E12.5), and the BM (E15.5) are also seeded by hematopoietic progenitors (Delassus and Cumano, 1996).

In the Absence of β1 Integrin Hematopoietic Progenitors are Generated Independently at Extra- and Intra-embryonic Sites and Accumulate in the Fetal Circulation

Since a complete knockout of the β1 integrin gene in mice results in peri-implantation lethality due to a defect in placental nidation (Stephens et al., 1995; Fässler and Meyer, 1995), it is impossible to assess directly the *in vivo* role of β1 integrins in hematopoiesis. To circumvent this obstacle somatic chimera were generated by injecting either β1$^{+/-}$ or β1$^{-/-}$ embryonic stem cells into C57/Bl6 wildtype blastocysts, allowing the development of chimeric mice after transfer into foster mothers. Presence and frequency of β1-null hematopoietic progenitors in the yolk sac, the p-Sp and fetal blood FB of chimeras were analysed at various stages during embryogenesis using clonogenic assays in methylcellulose or on stromal cells (Hirsch et al., 1996). To eliminate the wildtype-derived endogenous hematopoietic compartment, G418 was added to the cultures. The lack of the entire β1 integrin family did not affect the generation of hematopoietic progenitors in the yolk sac or the p-Sp. Prior to the onset of circulation (5-7 somite pairs) erythroid and myeloid potential was present both in the yolk sac and the PAS, whereas the potential to generate B cells was absent in the yolk sac and found exclusively at very low frequencies at the intra-embryonic site. After onset of circulation (8-12 somite pairs) lymphoid potential was encountered both in extra- and intra-embryonic sites. These data indicate that β1 integrin expression is not critical for the generation of early hematopoietic progenitors at intra- (p-SP) and extra-embryonic sites (yolk sac). Since no β1-null hematopoietic cells were found in fetal liver, thymus and spleen, β1 function is mandatory for the colonization of fetal sites of hematopoiesis. Most interestingly, β1-null hematopoietic progenitors are present in the fetal circulation.

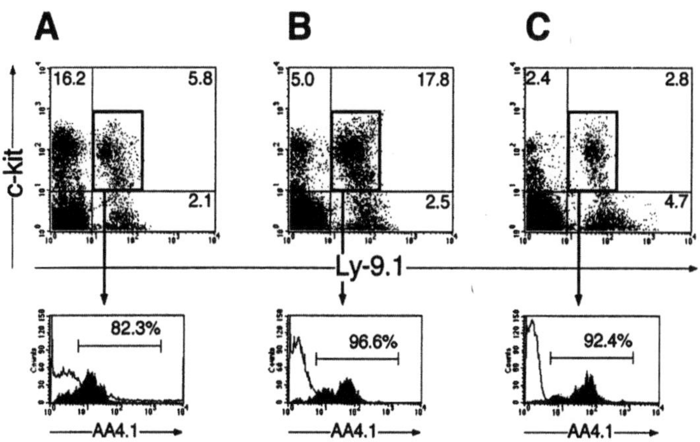

Fig. 2. Primitive hematopoietic progenitors lacking β1 integrin expression accumulate in the fetal circulation of chimeric embryos and express stem cell-specific surface antigens. Flow cytometry analysis of fetal blood obtained from E13.5 (**A**), E15.5 (**B**) and E18.5 (**C**) β1-null chimeric embryos for the expression of c-kit and Ly-9.1. At all timepoints analysed FB cells expressed similar percentages of the stem cell-specific antigen AA4.1 on c-kit$^+$ cells. Percentages for every population are depicted in the respective histogram.

At E15.5 the majority of β1-null cells in the chimeric embryos co-expressed the stem cell markers c-kit and AA4.1 and were negative for lineage-specific markers (Fig. 2B). This population of *bona fide* HSCs accumulated in the circulation of chimeric mice at a peak level at E15.5. In contrast to heterozygous chimeras, this stem cell-like population was still detectable in β1-null chimeras around birth (Fig. 2C).

Absence of β1 Integrin does not Block Hemato-lymphoid Differentiation *in vitro* or in Fetal Organ Cultures

To test whether the absence of β1 integrin interferes with hematopoiesis β1-null progenitors were isolated from E15.5 FB of chimeras and cultured on S17 stromal cells. Under these conditions FB cells - similar to p-Sp cells - developed into Ig secreting B cells. To investigate hematopoiesis in the context of the fetal liver microenvironment, we used a fetal liver reaggregation model to study the differentiation and survival of hematopoietic progenitors in an organ culture system. In this model, FB β1-null progenitors differentiated into erythroid, myeloid and B cells with similar frequencies as control cells. Similar to the efficient development of blood cells in fetal liver organ cultures, β1-null FB cells differentiated efficiently into αβ or γδ T cells in reaggregation thymus organ cultures. In summary, these data demonstrate that β1 integrins are essential for the seeding of fetal liver and thymus by fetal hematopoietic progenitors, but are not critical for hemato-lymphoid differentiation in those environments.

Fetal Blood β1-null Cells Fail to Differentiate into Lymphocytes

Since hematopoietic progenitors in the absence of β1 integrin are generated in the p-Sp - the site proposed to harbor the lymphoid potential - and show a virtually

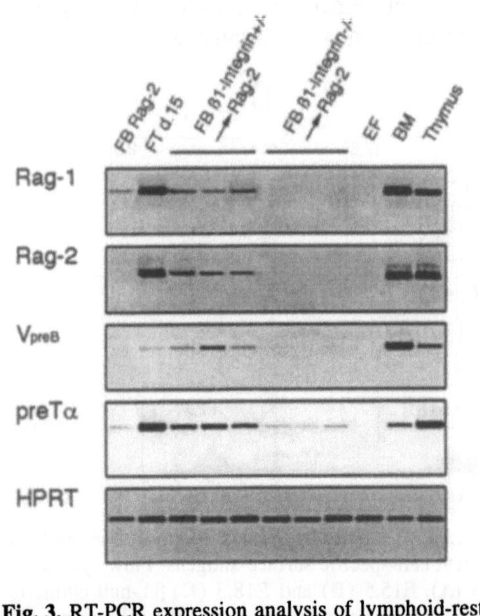

Fig. 3. RT-PCR expression analysis of lymphoid-restricted genes. Fetal blood (FB) was collected from E15.5 Rag-$2^{-/-}$ mice complemented either with $\beta1^{+/-}$ or $\beta1^{-/-}$ ES cells. Each lane corresponds to an individual embryo. Fetal thymus (FT), adult bone marrow (BM) and thymus served as positive controls; embryonic fibroblasts (EF) as negative control.

unaffected efficiency to undergo lymphopoiesis *in vitro*, we took advantage of the accumulation of β1-null cells in the fetal circulation to investigate the expression of lymphoid-restricted genes as a hallmark for lymphoid commitment *in vivo*. At no timepoint during embryogenesis nor in the adult chimeras did we observe expression of pan-lymphoid (Rag-2), B cell (λ5, Vpre B, Pax-5) or T cell (Gata-3) specific genes in β1-null hematopoietic cells isolated from yolk sac, AGM, FB (Fig. 3) or bone marrow of β1-null/Rag-2 chimeras.

These results strongly suggest that the development of B and T cells during embryogenesis requires the supportive microenvironment of the fetal liver or the fetal thymus respectively.

Concluding Remarks

It is currently unknown which cell adhesion receptors are responsible for the homing of HSCs to fetal liver, thymus, spleen and bone marrow during embryogenesis. The physiological role of some of these adhesion receptors was challenged by results obtained with genetically engineered mice lacking their expression. CD44-null mice showed no alterations in the generation of fetal liver HSCs and their efficiency to colonize the spleen, but did have an impairment in the egress of myeloid progenitors from the bone marrow (Schmits et al., 1997). Taking into consideration the normal development of hematopoiesis in these mice, these results did not support a critical role for CD44 in the migration of HSCs. In the case of integrins, the absence of α5 or α6 integrin still allowed

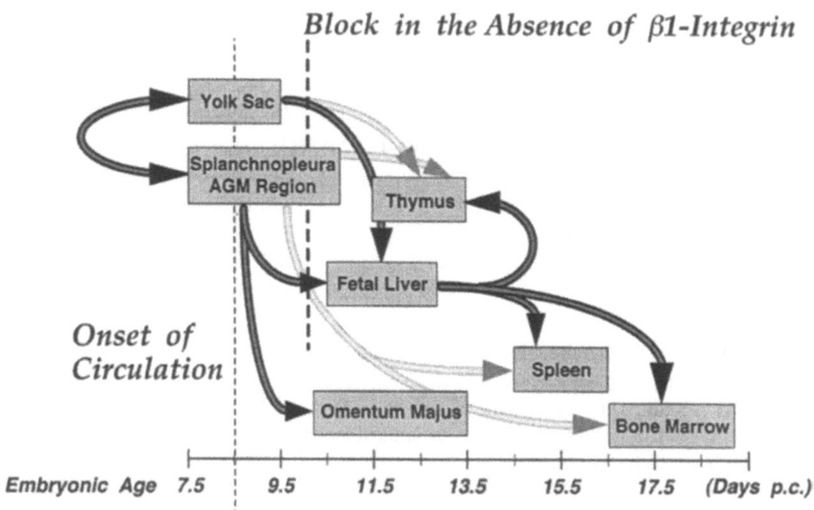

Fig. 4. Development of hematopoiesis in mouse embryogenesis. The different sites of hemato-lymphopoiesis are placed accordingly to their colonization during development. The main traits of migration of hematopoietic progenitors are depicted as dark grey arrows.

hemato-lymphoid differentiation which indicated that migration of hematopoietic progenitors to fetal liver, thymus and bone marrow was not abolished (Georges-Labouesse et al., 1996 and personal communication; Taverna et al., 1998). Using the Rag-2 complementation model, Arroyo et al. (1996) demonstrated that the absence of α4 integrin resulted in an early block of B-cell development in the BM and in an inability of thymic progenitors to leave the bone marrow. Remarkably, migration of α4-null HSCs into the fetal liver, spleen or bone marrow was not affected. Hence, α4 integrin is obviously dispensable for homing and seeding of hematopoietic organs by HSCs (Arroyo et al., 1999). Correspondingly, VCAM-1-null mice were found to have normal representation of all hematopoietic cells in thymus, spleen and bone marrow (Friedrich et al., 1996).

The absence of the β1 integrin family on hematopoietic cells blocks the colonization of the fetal liver by either yolk sac or p-Sp/AGM-derived HSCs (Fig. 4).

As mentioned above no single α chain mutation resulted in a corresponding phenotype. Therefore it might be either a combination of α subunits that can be deleted individually without effect or a yet unknown α subunit that heterodimerise with β1 integrin and mediate the functions essential for the homing of HSCs. An important task in the future will be to identify the α subunits that are important for colonisation of fetal sites of hematopoiesis, and to identify the integrin binding partners on the endothelial cells. Stem cell-specific knockouts of individual and groups of α subunits as well as endothelial cell-restricted ablation of genes encoding adhesion molecules should provide further information on the mechanism underlying the migration of hematopoietic cells.

Acknowledgements

The Basel Institute for Immunology was founded and is supported by F. Hoffmann-La Roche Ltd., Basel, Switzerland. The author would like to thank Dr. A. J. Young for critical reading of the manuscript.

References

Arroyo, AG, Yang, JT, Rayburn, H, Hynes, RO (1996) Differential requirements for α4 integrins during fetal and adult hematopoiesis. Cell 85, 997-1008

Arroyo, AG, Yang, JT, Rayburn, H, Hynes, RO (1999) α4 integrins regulate the proliferation/differentiation balance of multilineage hematopoietic progenitors *in vivo*. Immunity 11, 555-566

Belkin, AM, Zhidkova, I, Balzac, F, Altruda, F, Tomatis, D, Maier, A, Tarone, G, Koteliansky, VE, Burridge, K (1996) β1D integrin displaces the β1A isoform in striated muscles: localization at junctional structures and signaling potential in nonmuscle cells. J Cell Biol 132, 211-226

Chen, Q, Kinch, MS, Lin, TH, Burridge, K, Juliano, RL (1994) Integrin-mediated cell adhesion activates mitogen-activated protein kinases. J Biol Chem 269, 26602-26605

Chong, LD, Traynor-Kaplan, A, Bokoch, GM, Schwartz, MA (1994) The small GTP-binding protein Rho regulates a phosphatidylinositol 4-phosphate 5-kinase in mammalian cells. Cell 79, 507-513

Cumano, A, Dieterlen-Lièvre, F, Godin, I (1996) Lymphoid potential, probed before circulation in mouse, is restricted to caudal intraembryonic splanchnopleura. Cell 86, 907-916

Dedhar, S, Hannigan, GE (1996) Integrin cytoplasmic interactions and bidirectional transmembrane signalling. Curr Opin Cell Biol 8, 657-669

Delassus, S, Cumano, A (1996) Circulation of hematopoietic progenitors in the mouse embryo. Immunity 4, 97-106

Fässler, R, Georges-Labouesse, E, Hirsch, E (1996) Genetic analyses of integrin function in mice. Curr Opin Cell Biol 8, 641-646

Fässler, R, Meyer, M (1995) Consequences of lack of β1 integrin gene expression in mice. Genes Dev 9, 1896-1908

Friedrich, C, Cybulsky, MI, Gutierrez-Ramos, JC (1996) Vascular cell adhesion molecule-1 expression by hematopoiesis-supporting stromal cells is not essential for lymphoid or myeloid differentiation *in vivo* or *in vitro*. Eur J Immunol 26, 2773-2780

Georges-Labouesse, E, Messadeq, N, Yehia, G, Cadalbert, L, Dierich, A, Le Meur, M (1996) Absence of the α6 integrin leads to epidermolysis bullosa and neonatal death in mice. Nature Genet 13, 370-373

Godin, I, Dieterlen-Lièvre, F, Cumano, A (1995). Emergence of multipotent hemopoietic cells in the yolk sac and paraaortic splanchnopleura in mouse embryos, beginning at 8.5 days postcoitus. Proc Natl Acad Sci USA 92, 773-777

Hirsch, E, Iglesias, A, Potocnik, AJ, Hartmann, U, Fässler, R (1996) Impaired migration but not differentiation of haematopoietic stem cells in the absence of β1 integrins. Nature 380, 171-175

Houssaint, E (1981) Differentiation of the mouse hepatic primordium. II. Extrinsic origin of the haematopoietic cell lineage. Cell Differentiation 10, 243-252

Hynes, RO (1992) Integrins: versatility, modulation, and signaling in cell adhesion. Cell 69, 11-25

Hynes, RO (1996) Targeted mutations in cell adhesion genes: what have we learned from them? Dev Biol 180, 402-412

Keller, G, Lacaud, G, Robertson, S (1999) Development of the hematopoietic system in the mouse. Exp Hematol 27, 777-787

McNamee, HP, Ingber, DE, Schwartz, MA (1993) Adhesion to fibronectin stimulates inositol lipid synthesis and enhances PDGF-induced inositol lipid breakdown. J Cell Biol 121, 673-678

Medvinsky, AL, Samoylina, NL, Müller, AM, Dzierzak, EA (1993). An early pre-liver intraembryonic source of CFU-S in the developing mouse. Nature 364, 64-67

Medvinsky, A, Dzierzak, E (1996) Definitive hematopoiesis is autonomously initiated by the AGM region. Cell 86, 897-906

Russel E (1979) Hereditary anemias of the mouse: a review for geneticists. Adv Genet 20:357

Sabin FR (1920) Studies on the origin of blood vessels and of red corpuscles as seen in the living blastoderm of the chick during the second day of incubation. Contrib Embryol 9:213

Schlaepfer, DD, Hanks, SK, Hunter, T, van der Geer, P (1994) Integrin-mediated signal transduction linked to Ras pathway by GRB2binding to focal adhesion kinase. Nature 372, 786-791

Schmits, R, Filmus, J, Gerwin, N, Senaldi, G, Kiefer, F, Kundig, T, Wakeham, A, Shahinian, A, Catzavelos, C, Rak, J, Furlonger, C, Zakarian, A, Simard, JJ, Ohashi, PS, Paige, CJ, Gutierrez-Ramos, JC, Mak, TW (1997). CD44 regulates hematopoietic progenitor distribution, granuloma formation, and tumorigenicity. Blood 90, 2217-2233

Schwartz, MA, Lechene, C, Ingber, DE (1991) Insoluble fibronectin activates the Na/H antiporter by clustering and immobilizing integrin α5β1, independent of cell shape. Proc Nat Acad Sci USA 88, 7849-7853

Schwartz, MA, Schaller, MD, Ginsberg, MH (1995) Integrins: emerging paradigms of signal transduction. Annu Rev Cell Dev Biol 11, 549-599

Stephens, LE, Sutherland, AE, Klimanskaya, IV, Andrieux, A, Meneses, J, Pedersen, RA, Damsky, CH (1995) Deletion of β1 integrins in mice results in inner cell mass failure and periimplantation lethality. Genes Dev 9, 1883-1895

Taverna, D, Disatnik, MH, Rayburn, H, Bronson, RT, Yang, J, Rando, TA, Hynes, RO (1998). Dystrophic muscle in mice chimeric for expression of α5 integrin. J Cell Biol 143, 849-859

van der Flier, A, Kuikman, I, Baudoin, C, van der Neut, R, Sonnenberg, A (1995) A novel β1 integrin isoform produced by alternative splicing: unique expression in cardiac and skeletal muscle. FEBS Lett 369, 340-344

Vuori, K, Ruoslahti, E (1993) Activation of protein kinase C precedes α5β1 integrin-mediated cell spreading on fibronectin. J Biol Chem 268, 21459-21462

Wary, KK, Mainiero, F, Isakoff, SJ, Marcantonio, EE, Giancotti, FG (1996) The adaptor protein Shc couples a class of integrins to the control of cell cycle progression. Cell 87, 733-743

Zhidkova, NI, Belkin, AM, Mayne, R (1995) Novel isoform of β1 integrin expressed in skeletal and cardiac muscle. Biochem Biophys Res Commun 214, 279-285

The Ikaros Family and the Development of Early Intraembryonic Hematopoietic Stem Cells

J. Liippo, K.-P. Nera, P. Kohonen, M. Lampisuo, K. Koskela, P. Nieminen, and O. Lassila
Turku Graduate School of Biomedical Sciences, Department of Medical Microbiology, University of Turku, Kiinamyllynkatu 13, FIN-205020, Turku, Finland

Introduction

It is currently well established that across evolutionarily distant species the first blood-forming stem cells appear in the blood islands of both the extraembryonic yolk sac and the embryo proper. Following this initial and only transient generation of primitive erythroid-lineage cells, the major hematopoietic regions are thereafter located in the intraembryonic mesenchyme of the early embryos. Taking advantage of avian yolk sac-embryo chimeras it was originally shown that lymphoid cells are of intraembryonic origin [1, 2]. Two decades after the avian grafting experiments, the murine para-aortic splanchnopleura (P-Sp) and the aorta-gonad-mesonephros-region (AGM) were functionally characterized to have lymphoid potential [3, 4]. These results confirmed the intraembryonic sources for murine lymphopoiesis. Thus, mesodermal foci, i.e. the splanchnopleura and the ventrally derived intra-aortic clusters as well as the later appearing dorsal mesentery-derived para-aortic loci, initiate development of hematopoietic stem cells (HSCs). Although the AGM region is already known to harbor committed progenitors, and lymphoid potential has been shown to reside even in the para-aortic splanchnopleura, the exact time schedule of fate decisions is still rather controversial and unresolved. Contradictory evidence exists concerning the role of the P-Sp/AGM area on one hand as a de novo generator of intraembryonic hematopoietic stem cells and on the other hand as a site of hematopoietic differentiation. It has been postulated that these early hematopoietic environments only give rise to hematopoietic stem cells and further differentiation occurs in the lymphoid organs after colonization [5]. By contrast however, there are data showing the presence of committed lymphocyte progenitors not only in the fetal liver, but also in the murine AGM [6]. These results that are based on a clonal assay system support the idea of early lineage-determination in the lymphoid system. Taken together, these findings confirm not only the intraembryonic origin of the lymphoid cells, but also support the idea of early fate determination occurring before seeding of the primordial lymphoid organs.

Targeted disruptions of several transcription regulators have helped to gain new insights into the development of the embryonic hematopoietic system [7]. Although gene disruptions only reveal the primary role of a given transcription factor, they have played a central role in defining the complex regulatory network controlling lineage commitment and differentiation. Moreover, the traditional idea of gene regulators as mere transactivators has been challenged by novel findings showing that not only gene activation but also gene repression plays a crucial role in the control of gene expression within the hematopoietic system [8]. In addition, recent data demonstrate the intimate interplay between sequence-specific transcription factors and the globally acting chromatin remodellers that influence the accessibility of target gene loci [9]. Specifically, members of the Ikaros family are known to function at the interface between chromatin restructuring and cell cycle-specific as well as developmental gene regulation [10]. The Ikaros proteins are thought to function as local recruiters of the global chromatin remodeling factors that include the DNA-dependent ATPase Mi-2 as well as histone deacetylases [11].

To obtain better insight into the lymphoid potential of the early intraembryonic mesenchymal cells, we have assessed the role of the Ikaros family members in early avian hemato- and lymphopoietic development. We have in addition studied embryonic day (E) seven para-aortic precursors and their developmental potential. Using adoptive intrathymic cell transfers, transcription factor and TCR gene rearrangement analysis, we have shown that this progenitor cell population is capable of giving rise to T cell differentiation [12]. In line with the data demonstrating T and B lineage-restricted progenitors in the murine AGM and fetal liver and the presence of prebursally committed B cell precursors, our results further support the notion that diversification of lymphoid progeny occurs prior to the colonization of the lymphoid rudiments. Moreover, the early appearance of hematolymphoid transcription factors found in our studies is most likely a reflection of stem cell plasticity towards alternate lineage options.

Functional and Phenotypic Analysis of the Prethymic Progenitors in the Avian Intraembryonic Mesoderm

To assess the functional properties of the intraembryonic progenitors we conducted adoptive cell transfer experiments using prethymic cells isolated from the para-aortic region. Cells were sorted according to their surface antigen expression and were subsequently injected into the recipients' irradiated thymic lobes. Reconstitution efficiency was thereafter analyzed by following the expression of a congenic thymocyte marker. The results revealed that E7 progenitors expressing the cell surface antigens $\alpha2\beta1$-integrin, HEMCAM, c-kit and thrombomucin could recolonize the recipients' thymi [12]. These molecules are thought to have a role in the migration, homing and adhesion of the early progenitors. Thus, already prior to thymus seeding, the hematopoietic sites around the dorsal aortae contain T cell progenitors that can upon adoptive transfers give rise to functional T cells (Fig. 1). Nevertheless, it is difficult to conclude whether the E7 para-aortic cells are true T cell lineage-restricted progenitors or whether they represent a subset of hematopoietic stem cells.

According to our studies, the para-aortic progenitors retain their TCR genes in germline configuration. We have also demonstrated that the avian orthologs of GATA-3 and T cell factor-1 (Tcf-1) are both expressed early in ontogeny. The chicken Tcf-1 gene was found to generate isoforms that are differentially expressed during early embryogenesis and thymocyte development. This finding indicates a developmentally controlled expression of Tcf-1 splice variants. Taken together, these results demonstrate that although early progenitors express gene regulators related to the T cell lineage, they have not yet started to rearrange their antigen receptor genes.

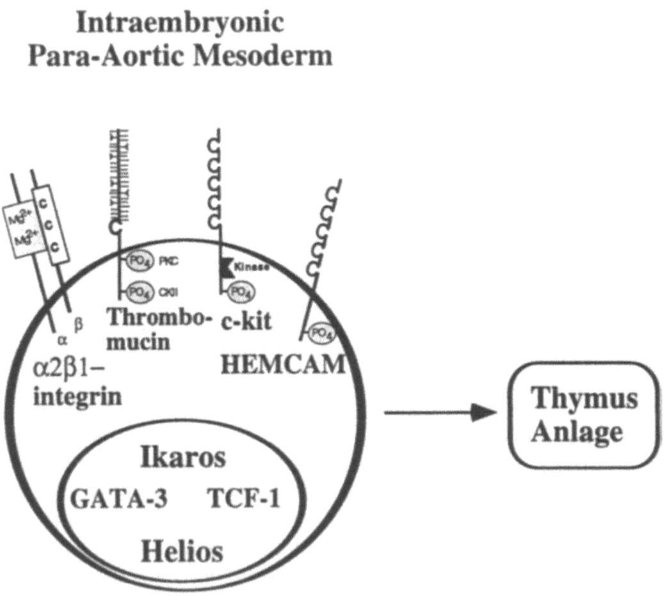

Fig. 1. Transcription factors and cell surface phenotype of the E7 intraembryonic para-aortic precursors capable of giving rise to T cell differentiation in the embryonic thymus. PKC, protein kinase C; CKII, casein kinase II; C, cysteine-rich domain; α, α2 chain; β, β1 chain.

Expression of the Ikaros Family Members in Early Avian Hematopoiesis and Lymphogenesis

Ikaros proteins comprise a family of chromatin remodeling-associated transcription factors that play a pivotal role in the regulation of lymphocyte development and function [10]. In contrast to earlier findings showing Ikaros and Aiolos as transactivators, they are also currently known to act as negative regulators and to serve as signaling thresholds in peripheral lymphocytes [8, 11]. According to the current view, Ikaros and Aiolos recruit histone deacetylases and thereby render specific gene loci into a condensed hypoacetylated state that is inaccessible to the transcriptional machinery. Moreover, these remodeling complexes that can be even 2MD in size enter the transcriptionally inert heterochromatin and also associate with DNA-replication foci. It is suggested that Ikaros and Aiolos thereby control the expression of lymphoid genes required for differentiation and lymphocyte function. The molecular mechanisms of Ikaros and Aiolos action become evident in mutant animals with targeted Ikaros or Aiolos alleles [10]. Ikaros-deficient animals with disrupted Ikaros DNA-binding domains fail to develop any lymphoid cells, whereas the corresponding heterozygous animals with reduced Ikaros activity develop lymphoproliferative disorders and tumors. The resulting developmental arrest and hyperproliferation arise due to the mutant non-DNA-binding Ikaros isoforms that exert a dominant-negative effect on other interacting proteins. Aiolos deficiency in turn results in profound aberrations specifically in B cell development and function. Loss of B cell tolerance, accelerated immunoglobulin production, B cell hyperproliferation and finally emergence of B cell tumors can be detected in Aiolos null animals. In conclusion, Ikaros and Aiolos seem to form a molecular break for the proliferation and chromosome propagation of lymphocytes, both of which play an important role in lymphocyte development and function.

Expression of the Avian Ikaros Gene Corresponds to Early Multilineage Gene Activation

Given the central role of Ikaros in lymphopoiesis we sought to identify the avian Ikaros homolog and analyse its expression in avian ontogeny. The results revealed that Ikaros transcripts were not only detected in the primordial lymphoid organs, i.e. the bursa of Fabricius and thymus, but were additionally found from E2 onwards both in the embryo proper and the extraembryonic yolk sac [13]. In addition, Ikaros transcripts were detected in the E7 para-aortic progenitors [12]. Although murine Ikaros has also been found in the early embryos and in the earliest hematopoietic stem cells, the exact molecular role of Ikaros in early development has remained unresolved. According to recent evidence, Ikaros-null HSCs display a reduced long-term-repopulation activity. Consistent with our own results showing concomitant expression of Ikaros and c-kit, it has been demonstrated that murine flk-2 and c-kit are expressed at lower levels in the Ikaros-mutant HSCs. Early Ikaros activity could correspond also to the appearance of hemangioblasts that are putative common precursors for both the endothelial and hematopoietic stem cells. Accordingly, it has been shown that GATA-3 is expressed in the chick allantoic bud possibly harboring hemangioblasts [14].

As hematopoietic stem cells give rise to the more committed lineages, they have been suggested to exhibit a multilineage gene expression program reflecting their developmental promiscuity. Multilineage gene activation has also been shown to precede the onset of lineage commitment and fate adoption in the hematopoietic system [15]. Restriction of the developmental capacity during hematopoietic development is therefore more likely to be a matter of gene repression rather than target gene activation. In this respect Ikaros proteins are good candidates as regulators of hemato- and lymphopoiesis-associated genes. Given their ability to direct the assembly of the multisubunit repressor complexes with histone deacetylation and chromatin remodeling capacity, it is evident that changes in chromatin structure are already intimately involved in the regulation of the earliest hematopoietic development.

The Avian Aiolos Gene is Expressed in the Bursa of Fabricius

As murine Aiolos is considered to be a key threshold-factor for the expression of differentiation- and growth-associated genes during the development and function of B lymphocytes, we have analysed the avian Aiolos homolog in the evolutionarily distinct chicken B lymphopoietic system. In contrast to Ikaros, Aiolos was not expressed in the early hematopoietic sites [16]. Transcripts were, however, detected in cells isolated from embryonic bursae on E12. Thereafter Aiolos expression is clearly found in developing bursal cells, and on E18 it peaks concomitantly with the highest frequency of IgM-positive cells found in the bursal follicles. Studies on murine Aiolos support this finding, as Aiolos expression is increased upon transition from pro-B cells to pre-B and mature B cells. Cells that were isolated from splenic germinal centers and the Harderian gland also expressed Aiolos transcripts. Although Aiolos was additionally expressed in reticulo-endothelial (RP) and Marek's disease virus (Cu) transformed B and T cell lines as well as in DT40 lymphoma cells, one Aiolos-negative B cell line was found. It however still remains to be seen, whether the Aiolos deficiency contributes to the malignant phenotype of this cell line. In conclusion, the evolutionarily conserved avian Aiolos gene is strongly expressed in B cells, supporting thus the important role of Aiolos in the development and function of B lymphocytes.

The Avian Helios Gene is Expressed in Early Ontogeny
In mammals, Helios, the third member of the Ikaros family, is expressed in the earliest hematopoietic stem cells. Later in development, its expression becomes restricted primarily to the T cell lineage [10]. In the absence of gene targeting results, function of the Helios gene remains unresolved. Helios is however likely to have repressive functions on target genes like Ikaros and Aiolos. Helios is known to interact and dimerize with other proteins within the Ikaros family. Similarly to the other family members, also the avian Helios gene is highly conserved in evolution. It is expressed already by the early intra- and extraembryonic cells from the first day of embryonic development onwards, i.e. even prior to the onset of Ikaros expression. Moreover, a phylogenic tree based on the currently known sequence information of the Ikaros family across species postulates that Helios could be the putative ancestral form of Ikaros and Aiolos. In conclusion, it can be suggested that Helios, together with Ikaros, is an important contributor to the earliest lineage-determining events that start to restrict the developmental potential of the hematopoietic stem cells.

Alternate Variants of the Ikaros Family in the Avian

The presence of multiple alternatively spliced transcription factor variants increases the plasticity of gene control by bringing more components to the gene regulatory complexes that are formed via highly orchestrated protein-protein interactions. Isoforms of the Ikaros family members are important recruiters of such large regulatory complexes that mediate chromosome propagation and nucleosome remodeling. In fact, across evolutionarily distant species the primary Ikaros transcript undergoes alternative splicing and produces multiple protein isoforms with different regulatory functions (Fig. 2 and [13, 10]). In mammals however, the Aiolos protein has been shown to exist in only one full-length form that contains all the exons encoding the four DNA-binding domains [17].

In order to assess the putative alternative splicing of the avian Aiolos transcripts, we conducted an RT-PCR assay that encompasses all the potentially spliced exons. Two amplification products of different size were repeatedly generated from various lymphoid tissues [16]. Further analyses showed that the smaller product lacked one of the central exons containing two of the four zinc finger domains used in DNA recognition. No differences were however detected as the expression of these two variants was analysed in various lymphoid tissues and during lymphocyte development. Importantly, this novel Aiolos isoform is structurally dominant negative, as it lacks more than one DNA-binding motif. Previous data show that Ikaros proteins with less than three N-terminal zinc fingers used in DNA recognition are incapable of DNA binding [10]. All such isoforms can therefore be also functionally dominant-negative, as they can still titrate by dimerization via the C-terminal zinc fingers intact regulators into transcriptionally inert complexes. It would be of great importance to further assess the role of this dominant-negative Aiolos isoform in the regulatory complexes influencing the accessibility and state of acetylation of lymphoid genes.

In addition to the new Aiolos isoform, there are Ikaros transcripts in the avian that have not previously been found in mammals. We and others have demonstrated that in the chicken and rainbow trout there is one isoform, Ik-1A, that selectively lacks the exon number six (unpublished observations). According to our results, this isoform is highly expressed and seems to be the predominant isoform in lymphoid tissues and developing lymphocytes. The function of this novel isoform is still unresolved. Interestingly, however, the sixth exon contains a module that is highly conserved across species and even across different Ikaros family members. Although this domain has not yet been functionally characterized, it corresponds to the same region where a 10-amino-acid in-frame deletion mutation in human Ikaros has been linked to childhood acute lymphoblastic leukemia [18]. In addition to this mutation, the leukemic cells express large amounts of dominant-negative non-DNA-binding isoforms. Given thus the known antioncogenic and repressive functions of Ikaros proteins, it can be suggested that this currently unknown domain mediates gene repression as it is disrupted in leukemic cells.

The larger number of Ikaros family isoforms in lower species has also been given further support by our recent results demonstrating the presence of at least four Helios isoforms in the chicken. In the mouse only two Helios variants have been found [10]. These variants are structurally similar to the Ikaros isoforms Ik-1 and Ik-2. Importantly, the novel forms of avian Helios arise due to alternate usage of a novel exon located in the middle of the region encoding the DNA-binding domains.

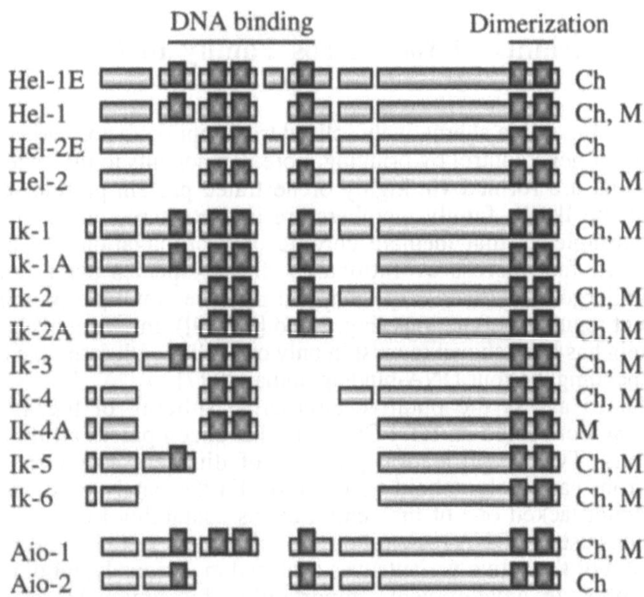

Fig. 2. Schematic illustration of the different alternatively spliced isoforms generated by the Ikaros family members in chicken (Ch) and mouse (M). The dark-colored perpendicular boxes indicate the zinc finger domains required for DNA binding and dimerization.

Concluding Remarks

The ability of stem cells to give rise to a number of oligolineage progenitor cells is a reflection of their initially versatile genetic program. This developmental plasticity is strikingly demonstrated through recent experiments that show the capacity of neuronal stem cells to transdifferentiate into HSCs [19, 20]. Moreover, also hematopoietic stem cells retain the ability to generate endothelial, muscle and neural cells [21]. In addition to the cell-intrinsic factors enabling adoption of alternate and even reciprocal fates, the multiplicity of various environmental cues is most likely intimately involved in regulating the developmental choices [20]. Whether stochastic or deterministic in nature, the lineage commitment or the exit from a stem cell state is eventually characterized by the appearance of progeny with restricted developmental capacity. Expression of key transcription factors at the stem cell level can still however be considered as a weak indication of commitment, as the developmental routes from embryonic stem cells via totipotent intermediates to tissue-forming stem cells are poorly understood. Therefore, the promiscuous multilineage gene activation that is present also in the early intraembryonic mesoderm harboring HSCs and hemangioblasts could only be considered as an indication of the potential to generate alternate fates.

Potential towards optional lineages becomes gradually diminished as the arising stem and progenitor cells start to adopt different developmental fates. Ikaros is considered to be a pivotal regulator active at the time when a fraction of the hematopoietic stem cells enters the lymphoid lineage. Aiolos in turn is thought to guide the progression of the developing B cell lineage. Given therefore the molecular role of Ikaros and Aiolos as chromatin remodeling-associated repressors, it can be proposed that their primary role is to restrict the developmental potential of stem and precursor cells towards more committed lineages. They dynamically recruit inappropriate genes to transcriptionally silent centromeric heterochromatin and allow the expression of genes required for the lineage commitment and lymphocyte function. Support for these hypotheses has been given by our own results showing the expression of Ikaros and Helios already in early ontogeny where the first lineage-determining decisions are taking place [12, 13]. Furthermore, identification of novel Ikaros family isoforms strongly suggests a more complex involvement of this family in the large chromatin remodeling complexes that most likely function already in early hematopoiesis [16]. Little is known however about the regulation and function of the different alternative isoforms. In line with the repressor hypothesis, it has been shown that Ikaros-based repression is lost in leukemic lymphocytes both in targeted animals and in humans with acute lymphoblastic leukemias [18]. Loss of repression is due to the predominance of oncogenic non-DNA-binding isoforms that prevent the function of other Ikaros family-associated factors via a dominant-negative effect. Although the novel Aiolos isoform that we have found structurally resembles these oncogenic Ikaros variants, there are currently no data showing the putative role of disrupted Aiolos function in human B cell malignancies. In conclusion, Ikaros-Aiolos-mediated gene repression plays a pivotal role not only in the restriction of the hematopoietic capacity during lymphocyte development, but also in suppressing the development of lymphoproliferative disorders.

Acknowledgements

Financial support for the present studies has been provided by the Academy of Finland, EU program BIOTECH (BIO4-CT97-2706), Special funds from the Turku University Central Hospital, the Finnish Medical Society Duodecim, the Finnish Medical Foundation, the Turku University Foundation, the Finnish Cancer Union, the Finnish Cultural Foundation of Southwest Finland and the Research and Science Foundation of Farmos. Ms M. Laine and A. Hakanen are acknowledged for their expert technical assistance.

References

1. Lassila O, Eskola J, Toivanen P, Martin C, Dieterlen-Lièvre F (1978) The origin of lymphoid stem cells studied in chicken yolk sac-embryo chimaeras. Nature 272:353-354
2. Lassila O, Martin C, Toivanen P, Dieterlen-Lièvre F (1982) Erythropoiesis and lymphopoiesis in chick yolk sac-embryo chimeras: contribution of yolk sac and intraembryonic stem cells. Blood 59:377-381
3. Medvinsky A, Dzierzak E (1996) Definitive hematopoiesis is autonomously initiated by the AGM region. Cell 86:897-906
4. Cumano A, Dieterlen-Lièvre F, Godin I (1996) Lymphoid potential, probed before circulation in mouse, is restricted to caudal intraembryonic splanchnopleura. Cell 86:907-916
5. Godin I, Garcia-Porrero JA, Dieterlen-Lièvre F, Cumano A (1999) Stem cell emergence and hemopoietic activity are incompatible in mouse intraembryonic sites. J Exp Med 190:43-52
6. Ohmura K, Kawamoto H, Fujimoto S, Ozaki S, Nakao K, Katsura Y (1999) Emergence of T, B, and myeloid lineage-committed as well as multipotent hemopoietic progenitors in the aorta-gonad-mesonephros region of day 10 fetuses of the mouse. J Immunol 163:4788-4795
7. Glimcher LH , Singh H (1999) Transcription factors in lymphocyte development - T and B cells get together. Cell 96:13-23
8. Koipally J, Renold A, Kim J, Georgopoulos K (1999) Repression by Ikaros and Aiolos is mediated through deacetylase complexes. EMBO J 18:3090-3100
9. Björklund S, Almouzni G, Davidson I, Nightingale KP, Weiss K (1999) Global transcription regulators of eucaryotes. Cell 96:759-767
10. Cortes M, Wong E, Koipally J, Georgopoulos K (1999) Control of lymphocyte development by the Ikaros gene family. Curr Opin Immunol 11:167-171
11. Kim J, Sif S, Jones B, Jackson A, Koipally J, Heller E, Winandy S, Viel A, Sawyer A, Ikeda T, Kingston R, Georgopoulos K (1999) Ikaros DNA-binding proteins direct formation of chromatin remodeling complexes in lymphocytes. Immunity 10:345-355
12. Lampisuo M, Liippo J, Vainio O, McNagny KM, Kulmala J, Lassila O (1999) Characterization of prethymic progenitors within the chicken embryo. Int Immunol 11:63-69
13. Liippo J, Lassila O (1997) Avian Ikaros gene is expressed early in embryogenesis. Eur J Immunol 27:1853-1857
14. Caprioli A, Jaffredo T, Gautier R, Dubourg C, Dieterlen-Lièvre F (1998) Blood-borne seeding by hematopoietic and endothelial precursors from the allantois. Proc Natl Acad Sci USA 95:1641-1646
15. Enver T, Greaves M (1998) Loops, lineage and leukemia. Cell 94:9-12
16. Liippo J, Mansikka A, Lassila O (1999) The evolutionarily conserved avian Aiolos gene encodes alternative isoforms. Eur J Immunol 29:2651-2657
17. Morgan B, Sun L, Avitahl N, Andrikopoulos K, Ikeda T, Gonzales E, Wu P, Neben S, Georgopoulos K (1997) Aiolos, a lymphoid restricted transcription factor that interacts with Ikaros to regulate lymphocyte differentiation. EMBO J 16:2004-2013.
18. Sun L, Heerema N, Crotty L, Wu X, Navara C, Vassilev A, Sensel M, Reaman GH, Uckun FM (1999) Expression of dominant-negative and mutant isoforms of the antileukemic transcription factor Ikaros in infant acute lymphoblastic leukemia. Proc Natl Acad Sci USA 96:680-685
19. Bjornson CR, Rietze RL, Reynolds BA, Magli MC, Vescovi AL (1999) Turning brain into blood: a hematopoietic fate adopted by adult neural stem cells in vivo. Science 283:534-537
20. Fuchs E, Segre JA (2000) Stem cells: a new lease on life. Cell 100:143-155
21. Weissman IL (2000) Stem cells: units of development, units of regeneration, and units in evolution. Cell 100:157-168

Function of Cytokines in Lymphocyte Development

M. Kondo and I. L. Weissman

Departments of Pathology and Developmental Biology, Stanford University School of Medicine, Stanford, CA 94305, USA

Introduction

All hematopoietic cells, including lymphocytes, are derived from hematopoietic stem cells (HSC) (Morrison et al. 1995). The process of hematopoiesis is initiated with an asymmetric stem cell division, giving rise to one cell that remains an HSC and another that begins to differentiate. Subsequent development leads to the loss of self-renewal activity and commitment to a specific hematopoietic lineage, thus restricting its differentiation potential. Cytokines have been shown to play important roles in hematopoiesis, regulating the viability, expansion and differentiation of developing cells (Watowich et al. 1996).

The cytokine receptor common γ (γ_c) chain is a shared component among the receptors for IL-2, IL-4, IL-7, IL-9 and IL-15 (Sugamura et al. 1996). Dysfunction of the γ_c chain causes severe combined immunodeficiency (X-SCID) in humans (Leonard 1996). Typical SCID patients have no T or NK cells. Targeted disruption of the γ_c gene in mice also results in severe lymphopenia, suggesting that the γ_c chain-related cytokines play an important role in lymphopoiesis (Cao et al. 1995; DiSanto et al. 1995; Ohbo et al. 1996). Studies into the role of specific cytokines in the γ_c family revealed that IL-7 and IL-15 are critical for lymphocyte development. IL-7- and IL-7Rα-deficient mice show a similar reduction in the number of T and B cells as is seen in γ_c chain knockout mice (Peschon et al. 1994; von Freeden-Jeffry et al. 1995). IL-15Rα knockout mice have no NK cells (Lodolce et al. 1998). The phenotype of these mice implies that IL-7 is the most important factor for supporting T and B cell development, and that IL-15 is an indispensable cytokine in NK cell development.

Here we show that different functions of γ_c chain-related cytokines regulate specific events during T, B and NK cell development. We also describe the isolation of common lymphoid progenitors (CLP) in mouse bone marrow, and briefly discuss the role of cytokine signaling in regulating the development from CLP to mature cell lineages.

Importance of Cytokines in Lymphopoiesis

γ_c chain knockout mice have very few T and B cells and have no detectable NK cells (Cao et al. 1995; DiSanto et al. 1995; Ohbo et al. 1996). Although one may assume that this phenotype is the result of a developmental block at the CLP stage, phenotypically similar immunocompromised mice including SCID mice and RAG1, RAG2 and Ikaros knockout mice have obstructions at different points of lymphoid developmental pathways (Weissman 1994). Furthermore, populations downstream of CLP (pro-T and pro-B cells) are observed in γ_c chain knockout mice, suggesting that the γ_c chain family cytokines may not act at the level of CLP, but instead are required for lymphocyte development after T and B lineage commitment (Kondo et al. 1997a). Specifically, although cellularity of the thymus in γ_c chain knockout mice is ~3% of wild type mice, the CD4/CD8 expression pattern is close to normal. In contrast, B cell development is almost completely blocked at the pro-B to pre-B transition point, where pre-B cell selection is required for the stage transition. These data suggest that different functions of the γ_c chain cytokine family are required in T and B cell development.

Below is a discussion of work accomplished by us and other groups that clearly demonstrates that T, B, and NK cells have unique requirements for the γ_c-cytokine family during development.

To address the role of γ_c family cytokines in cell survival, γ_c knockout mice were crossed with mice transgenic for the expression of the anti-apoptotic factor Bcl-2 (Kondo et al. 1997a). The number of blood T cells in γ_c chain knockout mice is ~10% compared to wild type mice. When crossed to *bcl-2* transgenic mice, T cell numbers in the blood increase six-fold. A similar rescue of T cell development by Bcl-2 is observed in the IL-7 and IL-7Rα knockout mice (Akashi et al. 1997; Maraskovsky et al. 1997). In addition to this increase in cell number, the high CD4/CD8 ratio seen in mature T cells of γ_c, IL-7 or IL-7Rα knockouts is returned to normal levels by the introduction of *bcl-2* transgenes (Akashi et al. 1997; Kondo et al. 1997a).

Further studies revealed the specific stages at which IL-7 impacts T cell development. IL-7 plays an important role in positive selection at the $CD4^+CD8^+$ (double-positive) stage in intrathymic T cell development (Akashi et al. 1997). During the positive selection, TCR^{lo} double-positive cells upregulate Bcl-2 and TCR levels, divide, and give rise to transitional intermediate progeny on the path to mature $CD4^+$ or $CD8^+$ single-positive cells (Akashi et al. 1998). Blocking the effect of IL-7 at positive selection prevents progression from the double-positive stage to the single-positive stage. Importantly, this developmental block is not seen in *bcl-2* transgenic mice, suggesting that the survival signal provided by IL-7 is indispensable for the transition to single-positive cells after positive selection. IL-7 is also necessary for the maintenance of Bcl-2 expression in peripheral T cells (Akashi et al. 1997). These data show that one of the major functions of the γ_c chain cytokine family in T cell development, especially IL-7, is to support cell viability.

Although the rescue of T cell development by Bcl-2 in γ_c, IL-7, and IL-7Rα knockout mice is impressive, T cell numbers are still only 50% compared to wild type mice. This suggests that the proliferation induced by γ_c-related cytokines is necessary for complete restoration of T cell development. In regard to this, von Freeden-Jeffry et al. showed that a smaller percentage of cells in the $CD3^-CD4^-CD8^-$ population in the thymus of IL-7 knockout mice are in cell cycle (von Freeden-Jeffry et al. 1997). Therefore, stimulation of $CD3^-CD4^-CD8^-$ cells by IL-7 and other γ_c chain-related cytokines is necessary for expanding the T cell pool to a normal level.

A slight increase in the number of B cells is observed in the blood of γ_c chain knockout mice with *bcl-2* transgenes (Kondo et al. 1997a). However, no significant difference is observed in the absolute number of pro-B, pre-B and surface IgM-positive B cell populations in the bone marrow of γ_c chain knockout mice and γ_c chain knockout mice expressing Bcl-2. Hence, the anti-apoptotic effect by γ_c chain-related cytokines, in this case IL-7, is not crucial for B cell development. IL-7 is known to be a major factor for expansion of the B cell pool by stimulating cell growth of immature B cells. Furthermore, almost all immature B cells in γ_c chain and IL-7Rα knockout mice do not express intracytoplasmic immunoglobulin μ heavy chain, which is necessary for the formation of a pre-B receptor with surrogate light chains (unpublished observation; [Corcoran et al. 1996]). This is likely the reason why B cell development is blocked at a pro-B/pre-B transition stage in γ_c chain knockout mice. A recent report shows that a mutation of the IL-7Rα gene in humans causes $T^-B^+NK^+$ immunodeficiency (Puel et al. 1998), suggesting that the role of IL-7 is different between mice and humans.

Impaired NK cell development in γ_c chain knockout mice is not rescued by enforced expression of Bcl-2 (Kondo et al. 1997a). The function of IL-15 in NK cell development is not yet clear.

Identification of the Earliest Lymphoid Lineage Cells, Common Lymphoid Progenitors in Mouse Bone Marrow

As mentioned above, IL-7 is one of the most important cytokines for lymphocyte development. The action of IL-7 is almost completely limited to the lymphoid lineage, suggesting that its receptor may be a marker for cells committed to the lymphoid lineage. Indeed, we identified a population of lymphoid-restricted progenitors in the lineage marker-negative, IL-7Rα-positive fraction of mouse bone marrow cells ($Lin^-IL\text{-}7R\alpha^+Thy\text{-}1^-Sca\text{-}1^{lo}c\text{-}Kit^{lo}$) (Figure 1, Kondo et al. 1997b). These cells give rise solely to lymphocytes at the clonal level and likely represent the most immature lymphoid lineage cell. We therefore refer to these cells as common lymphoid progenitors (CLP). The expansion potential of CLP is so vast that 2,000 intravenously injected CLP produce $>10^7$ T and B cells in RAG-2 knockout mice. Importantly, CLP do not give rise to any myeloid lineage cells. In methylcellulose culture, IL-7 is indispensable for inducing pro-B colony formation from CLP. Aside from this, however, we have not yet clarified the function of IL-7 in CLP. $Lin^-IL\text{-}7R\alpha^+Thy\text{-}1^-Sca\text{-}1^{lo}c\text{-}Kit^{lo}$ cells are present in IL-7 and γ_c chain knockout mice (unpublished observation). This suggests that IL-7 does not induce lymphoid lineage commitment in upstream populations. As mentioned above, IL-7 promotes cell survival by maintaining a high expression level of Bcl-2. However, Bcl-2 expression in CLP of IL-7 and γ_c chain knockout mice is comparable to wild type mice (unpublished observation), implying that Bcl-2 expression in CLP is regulated by other mechanisms.

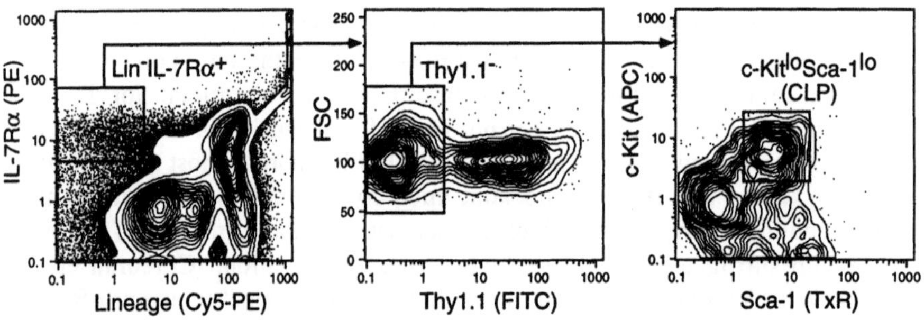

Fig. 1. Sorting procedure of the common lymphoid progenitors (CLP) from C57BL/Ka-Thy1.1 mice. Bone marrow cells are incubated with antibody cocktail for lineage markers (CD3, CD4, CD8, B220, Mac-1, Gr-1 and TER119). After depletion of lineage marker highly positive cells by magnetic beads, cells were stained with anti-rat IgG conjugated with Cy5-PE. After blocking with normal rat IgG, cells are stained with FITC-anti-Thy1.1, biotinylated anti-IL-7Rα, Texas Red-anti-Sca-1, and APC-anti-c-Kit. Cells are further incubated with streptavidin PE. Sorting gates are set first on lineage marker-negative (PI-negative) IL-7Rα-positive population (left). After setting the gate on $Thy1.1^-$ population (center), finally $c\text{-}Kit^{lo}Sca\text{-}1^{lo}$ population (right) is sorted by a flow-cytometer.

Recently, Nutt et al. showed that pro-B cells from *pax5* knockout mice can differentiate into myeloid and other lineage cells under certain conditions (Nutt et al. 1999). To initiate myeloid differentiation, IL-7 must be removed from the culture. One may hypothesize that CLP could differentiate into myeloid cells under myeloid cell culture conditions without IL-7. However, we have never observed CLP-derived myeloid colonies in methylcellulose culture in the presence of various cytokines (without IL-7) that support myeloid colony formation from hematopoietic stem cells (Kondo et al. 1997b). Also, *pax5* is unlikely to be responsible for preventing myeloid differentiation from CLP, since CLP express very low levels of *pax5* (unpublished observation). Importantly, CLP expressing exogenous human IL-2Rβ and both human GM-CSFRα and β_c subunits can differentiate into granulocytes and macrophages when stimulated with human IL-2 and human GM-CSF, respectively, *in vitro* (unpublished observation). This suggests that CLP may still have a latent myeloid lineage differentiation capability even after lymphoid lineage commitment. We are currently examining the molecular mechanisms that control the developmental outcomes in wild type and genetically altered CLP in an effort to define more precisely the process of lineage commitment.

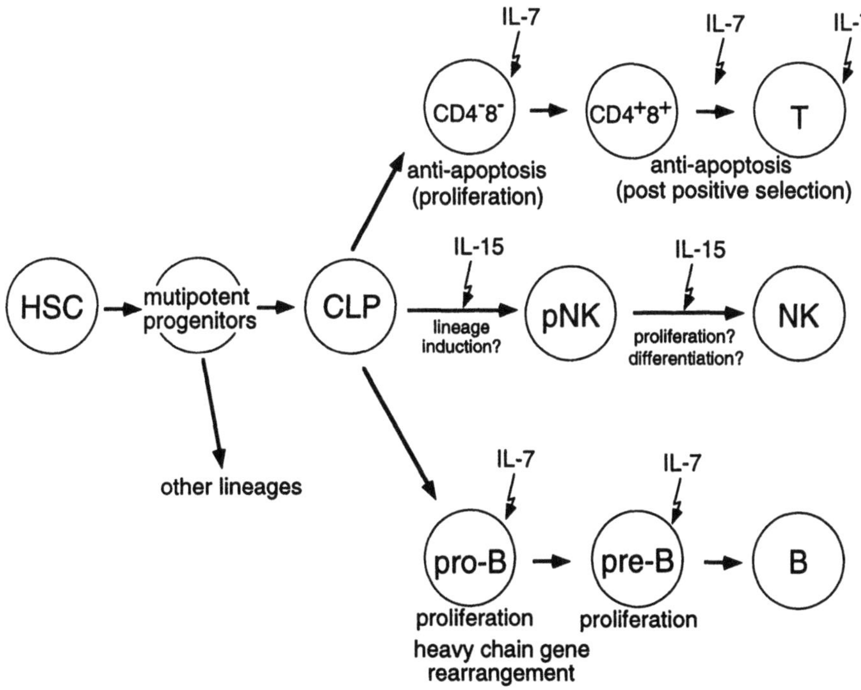

Fig. 2. Schematic summary of the function of IL-7 and IL-15 in lymphocyte development.

Conclusions

Cytokines play crucial roles in hematopoiesis and lymphopoiesis. The hypothesized role of cytokines has for the most part been restricted to factors that promote the proliferation of specific cell subsets. We now know that the function of cytokines is not limited to cell growth stimulation (Figure 2). For example, IL-7 promotes cell survival at several stages during T cell development. M-CSF has been shown to play a similar role in monocyte development (Lagasse and Weissman 1997). IL-7 also regulates specific developmental steps in B cells, being required for the rearrangement of the immunoglobulin heavy chain gene (Corcoran et al. 1998). It is important to investigate whether cytokines have instructive effects in lineage commitment. Some cytokines may specifically induce CLP from multipotent progenitors, or pro-T, pro-B or pro-NK cells from CLP. Further studies into the signal transduction pathways that drive the cellular response to cytokine stimulation will be of central importance in unveiling the molecular mechanisms of hemato-lymphopoiesis.

Acknowledgement

We are grateful to the colleagues in our laboratory for valuable comments. We would like to thank Libuse Jerabek for her excellent lab management, and Angela G King, David C Scherer and Dennise Dalma-Weiszhausz for critically reviewing the manuscript. This research was supported by USPHS grant 026 CA 42551. M.K. is a fellow of the Irvington Institute for Immunological Research.

References

Akashi K, Kondo M, von Freeden-Jeffry U, Murray R, Weissman IL (1997) Bcl-2 rescues T lymphopoiesis in interleukin-7 receptor-deficient mice. Cell 89: 1033-4

Akashi K, Kondo M, Weissman IL (1998) Two distinct pathways of positive selection for thymocytes. Proc Natl Acad Sci USA 95: 2486-91

Cao X, Shores EW, Hu-Li J, Anver MR, Kelsall BL, Russell SM, Drago J, Noguchi M, Grinberg A, Bloom ET et al. (1995) Defective lymphoid development in mice lacking expression of the common cytokine receptor gamma chain. Immunity 2: 223-38

Corcoran AE, Riddell A, Krooshoop D, Venkitaraman AR (1998) Impaired immunoglobulin gene rearrangement in mice lacking the IL-7 receptor. Nature 391: 904-7

Corcoran AE, Smart FM, Cowling RJ, Crompton T, Owen MJ, Venkitaraman AR (1996) The interleukin-7 receptor alpha chain transmits distinct signals for proliferation and differentiation during B lymphopoiesis. EMBO J 15: 1924-32

DiSanto JP, Muller W, Guy-Grand D, Fischer A, Rajewsky K (1995) Lymphoid development in mice with a targeted deletion of the interleukin 2 receptor gamma chain. Proc Natl Acad Sci USA 92: 377-81

Kondo M, Akashi K, Domen J, Sugamura K, Weissman IL (1997a) Bcl-2 rescues T lymphopoiesis, but not B or NK cell development, in common gamma chain-deficient mice. Immunity 7: 155-62

Kondo M, Weissman IL, Akashi K (1997b) Identification of clonogenic common lymphoid progenitors in mouse bone marrow. Cell 91: 661-72

Lagasse E, Weissman IL (1997) Enforced expression of Bcl-2 in monocytes rescues macrophages and partially reverses osteopetrosis in op/op mice. Cell 89: 1021-31

Leonard WJ (1996) The molecular basis of X-linked severe combined immunodeficiency: defective cytokine receptor signaling. Annu Rev Med 47: 229-39

Lodolce J P, Boone DL, Chai S, Swain RE, Dassopoulos T, Trettin S, Ma A (1998) IL-15 receptor maintains lymphoid homeostasis by supporting lymphocyte homing and proliferation. Immunity 9: 669-76

Maraskovsky E, O'Reilly LA, Teepe M, Corcoran LM, Peschon JJ, Strasser A (1997) Bcl-2 can rescue T lymphocyte development in interleukin-7 receptor-deficient mice but not in mutant rag-1-/- mice. Cell 89: 1011-9

Morrison SJ, Uchida N, Weissman IL (1995) The biology of hematopoietic stem cells. Annu Rev Cell Dev Biol 11: 35-71

Nutt SL, Heavey B, Rolink AG, Busslinger M (1999) Commitment to the B-lymphoid lineage depends on the transcription factor Pax5. Nature 401: 556-62

Ohbo K, Suda T, Hashiyama M, Mantani A, Ikebe M, Miyakawa K, Moriyama M, Nakamura M, Katsuki M, Takahashi K, Yamamura K, Sugamura K (1996) Modulation of hematopoiesis in mice with a truncated mutant of the interleukin-2 receptor gamma chain. Blood 87: 956-67

Peschon JJ, Morrissey PJ, Grabstein KH, Ramsdell FJ, Maraskovsky E, Gliniak BC, Park LS, Ziegler SF, Williams DE, Ware CB, et al. (1994) Early lymphocyte expansion is severely impaired in interleukin 7 receptor-deficient mice. J Exp Med 180: 1955-60

Puel A, Ziegler SF, Buckley RH, Leonard WJ (1998) Defective IL7R expression in T(-)B(+)NK(+) severe combined immunodeficiency. Nat Genet 20: 394-7

Sugamura K, Asao H, Kondo M, Tanaka N, Ishii N, Ohbo K, Nakamura M, Takeshita T (1996). The interleukin-2 receptor gamma chain: its role in the multiple cytokine receptor complexes and T cell development in XSCID. Annu Rev Immunol 14: 179-205

von Freeden-Jeffry U, Solvason N, Howard M, Murray R (1997) The earliest T lineage-committed cells depend on IL-7 for Bcl-2 expression and normal cell cycle progression. Immunity 7: 147-54

von Freeden-Jeffry U, Vieira P, Lucian LA, McNeil T, Burdach SE, Murray R (1995) Lymphopenia in interleukin (IL)-7 gene-deleted mice identifies IL-7 as a nonredundant cytokine. J Exp Med 181: 1519-2

Watowich SS, Wu H, Socolovsky M, Klingmuller U, Constantinescu SN, Lodish HF (1996). Cytokine receptor signal transduction and the control of hematopoietic cell development. Annu Rev Cell Dev Biol 12, 91-128

Weissman IL (1994) Stem cells, clonal progenitors, and commitment to the three lymphocyte lineages: T, B, and NK cells. Immunity 1, 529-31

Re-evaluation of B Lymphocyte Lineage Differentiation Schemes

P.W. Kincade, K.J. Payne, K.-S. Tudor, Y. Yamashita, K.L. Medina, M.I.D. Rossi and T. Kouro
Oklahoma Medical Research Foundation, Oklahoma City, OK 73104, USA

Introduction

Differentiation models for hematopoietic stem cells are to experimental hematologists what metabolic pathway charts are to biochemists. It has been extremely useful to consider the results of gene targeting and other experiments within the context of diagrams predicting key steps in B and T lymphopoiesis. Comparable schemes for human cells allow assignment of malignant cells to particular differentiation stages and interpretation of immunodeficiencies. Practical advances in multi-parameter flow cytometry and commercially available monoclonal antibodies further account for the popularity of differentiation schemes depicted on the walls of most laboratories. However, there is reason to believe that these models are not accurate or sufficiently complete in all respects. We will briefly recount how our recent studies forced re-evaluation of some widely accepted milestones for murine B lineage differentiation.

Down-regulation of CD43 and Other "Late" Events

Hardy and colleagues first determined that $sIgM^+$ B cells, and the largest population of small lymphocytes within bone marrow lack CD43, while presumptive early precursors are positive for this marker [1]. Virtually all $CD43^-$ cells synthesize the μ heavy chains of IgM, making CD43 "loss" one of the most popular milestones for the transition of pro-B to pre-B stages. Actually, the CD43 core protein continues to be expressed in newly formed B cells and it is down-regulation of a unique glycoform detected by the S7 monoclonal antibody that is informative. Furthermore, the S7 epitope is re expressed when mature B cells become activated [2].

We recently found a circumstance where very early B lineage lymphoid cells down-regulate the S7 epitope of CD43 [3]. A normally rare population of $CD19^+$ $BP\text{-}1^+$ $CD43^-$ cytoplasmic μ^- cells accumulate within bone marrow of estrogen treated normal, or $RAG^{-/-}$ mice. We initially thought the hormone caused "lineage progression" to the small pre-B cell stage. However, the cells continued to express the early marker TdT, along with Bcl-2 and an unusually large amount of their Ig genes remained in germ line configuration. Moreover, similar cμ$^-$ cells accumulated in marrow of mice bearing a rearranged human μ chain transgene on an immunodeficient RAG$^{-/-}$ background [3]. We were eventually forced to conclude that down-regulation of CD43 is not necessarily restricted to cells at the small pre-B cell stage. These and other studies revealed that elevated estrogen levels caused early pro-B cells to exit cell cycle and acquire some, but not all characteristics of pre-B cells [3].

The original definition of "pre-B" cells is based on expression of cytoplasmic μ chains and it is unfortunate that no pattern of surface markers precisely corresponds to the pro-B to pre-B cell transition. CD25/IL-2R is transiently acquired by many cells at that time, but is lost at later stages of differentiation [4]. The heparan sulfate proteoglycan, syndecan 4 is expressed by approximately half of the pro-B cells that can be cloned in IL-7 containing cultures and the marker is retained until B cells undergo isotype switching [5]. This, or some other proteoglycan is required for effective IL-7 recognition by pro-B cells [6]. Syndecan 4 is displayed at a slightly earlier stage than CD2 or CD72, two markers that do characterize most pre-B and B cells in the mouse [5 and P.W. Kincade, unpublished].

Low CD24 Expression as an "Early" Marker

It has always been clear that CD45R epitopes (often referred to as B220) are not restricted to the B lymphocyte lineage in mice [7]. However, they do serve as useful markers in some circumstances and particularly when used with additional antibodies. CD45R and CD24 represent a combination frequently used to distinguish "the earliest B lineage stage" in the mouse [8]. The original description utilized the 30F1 monoclonal antibody, but M1/69 is also commonly used in this manner [1, 9]. In our experience, the two reagents give different results when used with putative early precursors. We consider lineage (Lin) marker negative (that is CD45R$^-$) TdT$^+$ cells to be early pro-B cells, because they give rise to CD19$^+$ cells in culture (see below). More than half of these precursors have above background staining with the 30F1 antibody and even more are positive with M1/69 [10]. We think it would be difficult, if not impossible, to identify functional lymphocyte precursors at the Lin$^-$ stage on the basis of low levels of CD24 expression.

CD45R$^+$ CD19$^-$ Ly-6C$^+$ Lymphocytes

Early CD45R$^+$ cells should include precursors with the potential to express the more lineage restricted marker, CD19. However, CD45R$^+$ CD19$^-$ populations include a distinctive subset of NK lineage associated lymphocytes [11] that can be conveniently detected with NK1.1 or DX5 monoclonal antibodies. We found that an additional third of the CD45R$^+$ CD19$^-$ subset express Ly-6C and that none of the Ly-6C$^+$ lymphocytes yielded CD19$^+$ progeny when placed in culture [10]. In contrast, approximately one out of six CD45R$^+$ CD19$^-$ DX-5$^-$ Ly-6C$^-$ cells were functional B cell precursors and may represent major intermediates between stem cells and pre-B cells [10,12].

We found that the two subsets of the $CD45R^+$ $CD19^-$ category differ in two other important ways. The Ly-$6C^-$ lymphocytes expressed the IL-7Rα chain, consistent with their dependence on this cytokine, and lacked CD11b/Mac-1 [10]. In contrast, most of the Ly-$6C^+$ subset lacked IL-7Rα and were Mac-1^+. In addition, they were AA4.1^+ and a subset also displayed CD4. We assume that the $CD4^+$ category corresponds to a population of lymphocytes previously noted by Rolink and colleagues to lack functional B cell precursor activity [11]. In any case, these characteristics of Ly-$6C^+$ lymphocytes closely match those described by Hardy and colleagues for cells designated fraction "A" [13].

The origin and fate of Ly-$6C^+$ cells is unclear because no one has demonstrated that they are efficient B cell precursors. We recently found that they persist in bone marrow of estrogen treated mice, although Lin^-TdT^+ early pro-B cells are suppressed [3]. This again questions their position in the major pathway of B lymphopoiesis. They might be defective cells that are destined to die, they could be non-functional as a result of the isolation procedure, or they could be destined to give rise to dendritic or T lineage lymphocytes. Until these issues are settled, it does not seem appropriate to refer to Ly-$6C^+$ cells as "the earliest B lineage stage".

TdT as a Marker for Early Lymphocyte Precursors

Osmond and colleagues identified lymphocytes in murine bone marrow that they designated "early pro-B cells" on the basis of their expression of the nuclear enzyme terminal deoxynucleotidyl transferase (TdT) and lack of CD45R [14]. Their staining procedure required fixation and permeabilization of cells, so the differentiation potential of early pro-B cells was not formally demonstrated. We recently found that Lin^-TdT^+ cells have distinctive surface properties and viable cells enriched on that basis have the potential to give rise to $CD19^+$ cells in single cell cultures [10,15].

Display of c-kit as a Distinction for Pro-B Cells

Relatively high levels of the transmembrane tyrosine kinase type receptor for stem cell factor, c-kit, is expressed by stem cells and myeloid progenitor cells in the mouse. This marker has also been extensively used to enrich B lineage cells with the potential to expand and differentiate in culture [16]. We found that Lin^- TdT^+ cells have a distinctive low density of c-kit [15]. Most of the B cell precursor activity was identified on this basis and when this was used alone as a sort criterion the resulting cells were up to 30% TdT^+. Thus, we confirmed that c-kit is a useful marker for early cells with the potential for generating $CD19^+$ lymphocytes. However, a minor population of Lin^-TdT^+ cells was identified that totally lacked c-kit [15]. Their rapid differentiation in culture suggested that they must be further along in the pathway than c-kit^{lo} precursors. Therefore, while c-kit provides a useful means for enriching $Lin^{-}TdT^{+}$ cells a subset of slightly more differentiated cells would be discarded on this basis. Interestingly, some of the Lin^{-} c-kit^{-} cells transiently re-expressed this growth factor receptor when placed in culture [15]. We also concluded that c-kit is down-regulated by a majority of B lineage cells soon after their acquisition of CD45R. This followed from two experimental findings. First, many $CD45R^{+}$ $CD43^{+}$ BP-1^{-} $sIgM^{-}$ cells lacked c-kit and retained the ability to give rise to $CD19^{+}$ lymphocytes. In addition, c-kit was absent from many $CD45R^{+}$ cells that we generated in short term cultures from Lin^{-} populations.

Nomenclature for Murine B Lineage Cells

Much confusion in this field arises from the lack of consistently used terminology. Some designate stages of B lineage differentiation on the basis of immunoglobulin gene rearrangement status [16]. While this has theoretical merit, there are no ideal combinations of surface markers that allow their isolation as viable purified cells. Those who study T lymphocyte lineage differentiation in the thymus avoided confusion by referring to surface marker characteristics. That is, cells lacking CD4 and CD8 are designated "double negative" and further subsets (DN I to DN IV) are resolved on the basis of CD25 and CD44 expression. It is presumably appreciated by all workers in that field that genetic polymorphisms can influence levels of CD44 expression in different strains of mice [17]. Technical aspects such as the epitope specificity of different monoclonal antibodies, as well as the degree and choice of fluorochrome labeling could also contribute to lab-to-lab variability. The same considerations apply to those who study stages of B lymphopoiesis, or other pathways of blood cell differentiation. In a perfect situation, cells would be given B lineage precursor assignment only if they have the potential to give rise to $CD19^{+}$ cells. However, it has to be acknowledged that many defective, but still interesting cells are generated as a consequence of the Ig rearrangement process and they might be proven to have distinctive surface properties [18].

We have not found a perfect solution to the nomenclature problem and rely primarily on terms introduced by Osmond and colleagues [14]. Kondo and colleagues recently introduced the term "common lymphoid progenitor" to describe a long suspected category of "lymphoid stem cells" [19]. Their choice of terms seems appropriate because the cells they isolated did not have extensive self-renewal capability after transplantation. They used IL-7Rα, c-kit and Ly-6A/Sca-1 as sorting parameters to obtain Lin^{-} cells with the potential to produce B and T lymphocytes, but not myeloid cells. It is not clear if all stem cell progeny traverse a stage with these characteristics en route to becoming pro-B cells and thymocytes. Common lymphoid progenitors share many surface features with cells we designated "early pro-B cells" [10,12] in accordance with Osmond's use of that term [14]. Studies are underway to determine if early pro-B cells efficiently give rise to T lymphocyte lineage cells.

Osmond and colleagues resolved $CD45R^{+}$ TdT^{+} cells in bone marrow and designated them "intermediate pro-B cells". This seems an appropriate term for $CD45R^{+}$ $CD19^{-}$ Ly-$6C^{-}$ NK1.1/DX-5^{-} lymphocytes that quickly yield $CD19^{+}$ cells in culture [10]. As discussed above, these cells express CD24 and therefore would not be classified as "fraction A". They also display the IL-7Rα chain and lack CD11b/Mac-1.

It is difficult to explicitly use the "late pro-B cell" terminology as Osmond did for $CD45R^+$ TdT^- cells, because this would include NK lineage precursors and "fraction A". It appears unlikely that either of those subsets are in a direct pathway between stem cells and pre-B cells. We have provisionally used the "late pro-B" term for $CD19^+$ $c\mu^-$ cells, but it remains to be seen how many of them continue to express TdT. Regardless, there would be little disagreement that $c\mu^+$ $sIgM^-$ cells should be referred to as "pre-B cells" that can be resolved into large, mitotically active and small non-dividing precursors.

Concluding Remarks

We have attempted to stress in our primary publications that B lineage precursors are not entirely homogenous with respect to the markers available to characterize them. Workers in the field are familiar with mouse strain dependent differences that further confound use of some monoclonal antibodies and there is no widely accepted nomenclature. More substantive issues relate to the nature of early lymphocyte precursors, where cells with unknown fate may not be the ones of most interest. These are all problems that can and should be solved so that we have a firm basis for understanding the many genetically engineered mice that are being produced. More importantly, we will identify early control points that are regulated by hormones and other components of the bone marrow environment.

Acknowledgements

Our work is supported by grants AI 20069 and AI 33085 from the National Institutes of Health. P.W.K. holds the William H. and Rita Bell Chair in biomedical research.

References

1. Hardy RR, Carmack CE, Shinton SA, Kemp JD, Hayakawa K (1991) Resolution and characterization of pro-B and pre-pro-B cell stages in normal mouse bone marrow. J Exp Med 173:1213-1225
2. Gulley ML, Ogata LC, Thorson JA, Dailey MO, Kemp JD (1988) Identification of a murine Pan-T cell antigen which is also expressed during the terminal phases of B cell differentiation. J Immunol 140:3751-3757
3. Medina KL, Strasser A, Kincade PW (2000) Estrogen influences the differentiation, proliferation, and survival of early B-lineage precursors. Blood 95:2059-2067.

4. Rolink A, Grawunder U, Winkler TH, Karasuyama H, Melchers F (1994) IL-2 receptor α chain (CD25, TAC) expression defines a crucial stage in pre-B cell development. Int Immunol 6:1257-1264
5. Yamashita Y, Oritani K, Miyoshi EK, Wall R, Bernfield M, Kincade PW (1999) Syndecan-4 is expressed by B lineage lymphocytes and can transmit a signal for formation of dendritic processes. J Immunol 162:5940-5948
6. Borghesi LA, Yamashita Y, Kincade PW (1999) Heparan sulfate proteoglycans mediate interleukin-7-dependent B lymphopoiesis. Blood 93:140-148
7. Scheid MP, Landreth KS, Tung JS, Kincade PW (1982) Preferential but nonexclusive expression of macromolecular antigens on B-lineage cells. Immunol Rev 69:141-159
8. Li YS, Wasserman R, Hayakawa K, Hardy RR (1996) Identification of the earliest B lineage stage in mouse bone marrow. Immunity 5:527-535
9. Hunte B.E., Capone M., Zlotnik A., Rennick D., Moore T.A. (1998) Acquisition of CD24 expression by $Lin^{-}CD43^{+}B220^{low}ckit^{hi}$ cells coincides with commitment to the B cell lineage. Eur J Immunol 28:3850-3856
10. Tudor K-S, Payne KJ, Yamashita Y, Kincade PW (2000) Functional assessment of precursors from murine bone marrow suggests a sequence of early B-lineage differentiation events. Immunity (in press)
11. Rolink A, Ten Boekel E, Melchers F, Fearon DT, Krop I, Andersson J (1996) A subpopulation of $B220^{+}$ cells in murine bone marrow does not express CD19 and contains natural killer cell progenitors. J Exp Med 183:187-194
12. Kincade PW, Medina KL, Payne KJ, Rossi MID, Tudor K-S, Yamashita Y, Kouro T (2000) Early B lymphocyte precursors and their regulation by sex steroids. Immunol Rev 175: In Press
13. Allman D, Li J, Hardy RR (1999) Commitment to the B lymphoid lineage occurs before D_H-J_H recombination. J Exp Med 189:735-740
14. Osmond DG (1990) B cell development in the bone marrow. Semin Immunol 2:173-180
15. Payne KJ, Medina KL, Kincade PW (1999) Loss of c-kit accompanies B-lineage commitment and acquisition of CD45R by most murine B-lymphocyte precursors. Blood 94:1-12
16. Rolink A, Melchers F (1993) Generation and regeneration of cells of the B-lymphocyte lineage. Curr Opin Immunol 5:207-217
17. Haegel H, Ceredig R (1991) Transcripts encoding mouse CD44 (Pgp-1,Ly-24) antigen: strain variation and induction by mitogen. Eur J Immunol 21:1549-1553
18. Ehlich A, Martin V, Muller W, Rajewsky K (1994) Analysis of the B-cell progenitor compartment at the level of single cells. Curr Biol 4:573-583
19. Kondo M, Weissman IL, Akashi K (1997) Identification of clonogenic common lymphoid progenitors in mouse bone marrow. Cell 91:661-672

II

Vessel Development

Growth Factors Regulating Lymphatic Vessels

A. Lymboussaki[1], M.G. Achen[2], S.A. Stacker[2] and K. Alitalo[1]
[1]Molecular/Cancer Biology Laboratory, Haartman Institute, PL21 (Haartmaninkatu), 00014 University of Helsinki, Finland
[2]Ludwig Institute for Cancer Research, Post Office Box 2008, Royal Melbourne Hospital, Victoria 3050, Australia

Introduction

The lymphatic vessels penetrate tissues of the body as a dense network that controls the microcirculation by draining fluid from the interstitial spaces. This fluid, lymph, is filtered by lymph nodes and returns to the systemic circulation through the thoracic and lymphatic ducts and the lymphaticovenous anastomoses [1]. The lymphatic endothelium differs from the blood vascular endothelium in many morphological aspects. Lymphatics have an irregular and wide lumen, an attenuated endothelial wall and, in general, do not develop a continuous basement membrane [2]. Lymphatic endothelial cells lack tight junctions and possess anchoring filaments attaching them to the surrounding connective tissue [1]. Due to their highly permeable nature and poorly developed basement membranes, the lymphatics have been proposed as the main routes for tumor metastasis [3]. Abnormal function of the lymphatics is implicated in disease states such as lymphedema, inflammation, infectious and immune diseases, fibrosis and tumors such as Kaposi´s sarcoma and lymphangioma/lymphangiomatosis [4].

Lymphatic vessels develop after the blood circulation begins and derive from embryonic veins [5]. Adult lymphatics can regenerate by sprouting from pre-existing lymphatic vessels. Similar mechanisms operate in lymphangiogenesis and angiogenesis, such as sprouting, intussusceptive and intercalated growth, fusion and regression [6]. The occurrence of lymphangiogenesis in solid tumors is a matter of debate. Intra-tumoral lymphatics have been identified but it is not clear whether they are residual, newly formed, obstructed, malformed, or labyrinthine [7]. The confusion regarding lymphangiogenesis in cancer partly reflects the lack of specific markers for the lymphatic endothelium.

Proteins that Signal for Lymphangiogenesis

The protein growth factors thought to control lymphangiogenesis, vascular endothelial growth factor-C (VEGF-C) and VEGF-D, are members of the VEGF family (for review see [8]) that consists of structurally related, dimeric glycoproteins sharing functional characteristics. Intra-chain disulphide bonds, forming the "cystine knot motif" [9], are conserved features of these proteins. These growth factors activate endothelial cell-specific VEGF receptors (VEGFRs) triggering signal transduction. Of the VEGFRs, VEGFR-3 (Flt4) has been implicated in the regulation of lymphatic development and lymphangiogenesis.

VEGF-C

VEGF-C was the first protein shown to be lymphangiogenic [10]. Full-length VEGF-C consists of an N-terminal pro-peptide, a VEGF-homology domain that is highly conserved among all VEGF family members, and a C-terminal pro-peptide [11]. VEGF-C is first synthesized as a 58 kDa precursor, which undergoes dimerization before secretion into the culture medium. Stepwise proteolytic processing of VEGF-C generates several forms with increased binding and activity towards VEGFR-3 and VEGFR-2, as is the case for VEGF-D (Fig. 1) [12]. VEGFR-3 is expressed by lymphatic endothelial cells in adult tissues [13] and is thought to signal for lymphangiogenesis [14], whereas VEGFR-2 (Flk1/KDR), a receptor for VEGF, is broadly expressed on endothelial cells throughout embryonic development [15]. Non-processed VEGF-C preferentially binds and activates VEGFR-3, whereas the fully processed mature form, consisting of dimers of the VEGF homology domain, is a high affinity ligand for both VEGFR-3 and VEGFR-2 [12]. A VEGF-C mutant has been generated that activates VEGFR-3, but not VEGFR-2 [16].

The pattern of *VEGF-C* gene expression during embryogenesis suggests that VEGF-C plays a role in the development of the lymphatics. In the mouse, a paracrine relationship is suggested for VEGF-C and the lymphatic receptor, VEGFR-3, as transcripts for VEGF-C are located in mesenchymal cells adjacent to VEGFR-3 positive endothelia. VEGF-C transcripts are localised in the mesenchyme around large sac-like structures in the jugular area, and VEGFR-3 is found lining the borders of these sacs [17]. It is thought that the first lymphatic vessels sprout from such venous sac-like structures [18]. Accordingly, VEGFR-3 is present in the endothelial cells lining these vessels. VEGF-C is also found in the mesenterial connective tissue and VEGFR-3 is detected on developing lymphatic vessels at this site [17].

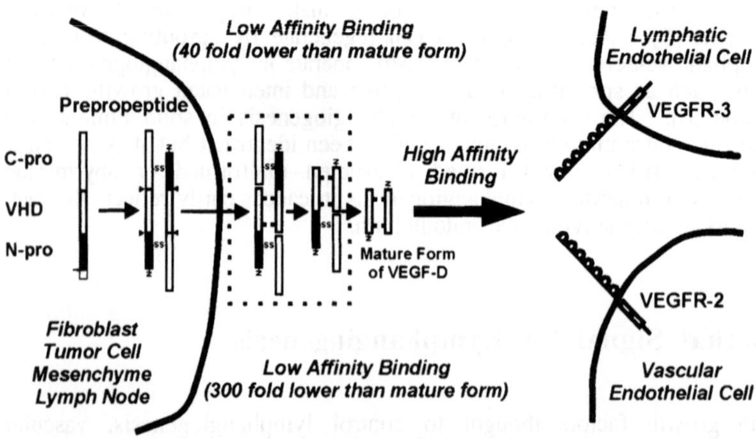

Fig. 1. Schematic representation of the proteolytic processing of the VEGF-C/VEGF-D proteins. These growth factors are initially synthesized as preproproteins containing N- and C-terminal propeptides (N-pro and C-pro) and a VEGF homology domain (VHD). Stepwise proteolytic processing generates several forms with increased binding and activity towards VEGFR-3 and VEGFR-2. The mature forms, consisting of dimers of the VEGF homology domain, bind the receptors with highest affinity. The processing shown is that for VEGF-D [25] which is similar to that for VEGF-C [12]. Note that VEGFR-2 is also expressed by at least some lymphatic vessels [44] and that VEGFR-3 can be detected in angiogenic blood vessels in tumors [37].

Two approaches have demonstrated that VEGF-C is lymphangiogenic. Firstly, overexpression of VEGF-C in the skin of transgenic mice resulted in lymphatic, but not vascular, endothelial proliferation and vessel enlargement [19]. Secondly, application of VEGF-C to the mature chick chorioallantoic membrane resulted in robust lymphangiogenic, but weak angiogenic, responses [10]. However, VEGF-C was mitogenic and chemotactic for vascular endothelial cells *in vitro* [20,21]. Furthermore, VEGF-C stimulated angiogenesis *in vivo* in the setting of limb ischemia [21] and in limbal vessels in the mouse cornea [22]. Thus, VEGF-C can regulate physiological and pathological blood vessel growth. It is possible that different lymphangiogenic/angiogenic responses occur depending on whether VEGF-C is proteolytically processed, whether VEGFR-3 is expressed on the endothelial cells of the blood vasculature, as is the case in early embryogenesis [17] and in tumors [23], and on other as yet ill-defined factors. In addition to lymphangiogenic and angiogenic properties, VEGF-C increases vascular permeability [12].

VEGF-D

VEGF-D is closely related in structure to VEGF-C as they share 48% amino acid sequence identity and have similar N- and C-terminal propeptides, a characteristic which distinguishes them from other members of the VEGF family [24]. VEGF-D is initially synthesized as a preproprotein that is proteolytically processed in a similar fashion to VEGF-C (Fig. 1) [25]. Complete processing gives rise to a mature form consisting of dimers of the VEGF homology domain. Cleavage of the propeptides from the VEGF homology domain occurs outside the cell [25]. The mature form of VEGF-D binds and activates both VEGFR-2 and VEGFR-3 [24]. Proteolytic processing of VEGF-D modulates receptor affinity as unprocessed VEGF-D binds VEGFR-2 and VEGFR-3 with much weaker affinity than does the mature form. Complete processing increases the affinity of VEGF-D for VEGFR-3 40-fold and for VEGFR-2 270-fold [25]. Mature VEGF-D exists predominantly in the form of a non-covalent dimer [25], as does VEGF-C [12].

In mouse embryos, the strongest sites of *VEGF-D* gene expression are in the lung mesenchyme (Fig. 2) and in a region immediately under the skin that is rich in developing melanocytes and fibroblasts [25,26]. In the skin, the position of the signal for VEGF-D transcripts is immediately adjacent to a layer of cells expressing VEGFR-3 (Achen *et al.*, unpublished) suggesting that VEGF-D may play a role in attracting the growth of the lymphatic vessels into the skin. The expression patterns for VEGF-D and VEGF-C overlap in many tissues as transcripts for both were detected around the developing metanephros and in the nasopharyngeal, jugular and neck areas in a manner complementary to VEGFR-3 ([27], Lymboussaki *et al.*, unpublished). However, the VEGF-D expression pattern was distinct from that for VEGF-C in the developing lung and the endocardial cushion tissue (Lymboussaki *et al.*, unpublished).

Expression of the *VEGF-D* gene is up-regulated by the nuclear oncogene c-fos [28]. Tumor cells lacking c-fos fail to become malignant, possibly due to an inability to induce angiogenesis [29], suggesting the existence of angiogenic factor(s) up-regulated by c-fos during tumor progression. Therefore, VEGF-D may play a role in c-fos-mediated angiogenesis during tumor progression.

VEGF-D is mitogenic for bovine aortic endothelial cells but is about five-fold less potent than $VEGF_{164}$ [24] and of similar potency to VEGF-C. Treatment of human umbilical vein endothelial cells with VEGF-D induces cell growth, elongation, branching and formation of a network of capillary-like cords [30]. Furthermore, VEGF-D is angiogenic in the rabbit cornea *in vivo* [30]. The lymphangiogenic capacity of VEGF-D is currently under investigation, but it is anticipated that VEGF-D may be lymphangiogenic in a similar fashion to VEGF-C,

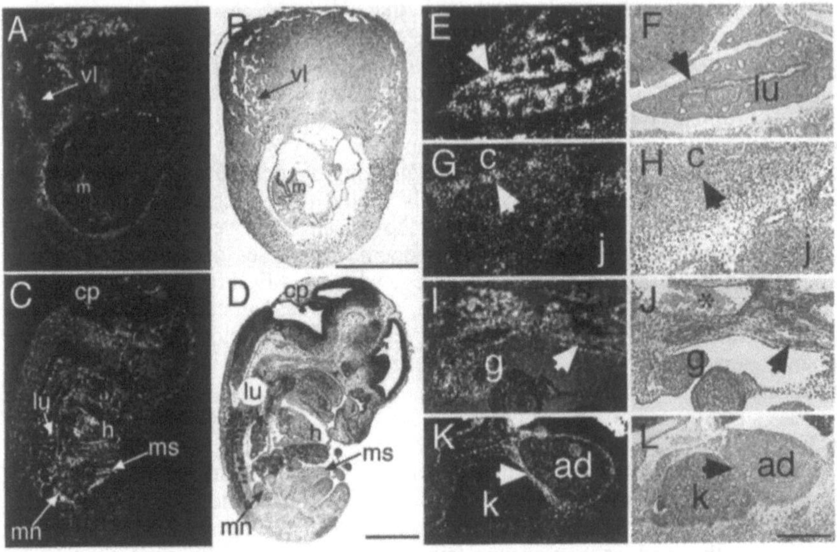

Fig. 2. Localization of VEGF-D transcripts in mouse embryos analyzed by *in situ* hybridization. A. E8.5 embryo: VEGF-D is expressed in the head mesenchyme (m) and in some cells surrounding the venous lacunae (vl). C. E12.5 embryo: VEGF-D transcripts are in the endocardial cushion tissue of the heart (h), the metanephric area (mn) and the mesenterium (ms). E-L. Higher power of regions expressing abundant VEGF-D transcripts. E. E13.5 embryo: VEGF-D transcripts (arrow) in the lung (lu). G. E13.5 embryo: signal (arrow) in the jugular area (j). I. E14.5 embryo: VEGF-D in the mesenchyme in close proximity to the mesenterial vessels (arrow). Blood cells gave a false positive signal (see asterisk in J). K. The linings of the adrenal gland (ad) and the kidney (k) are strongly positive (arrow) at E15.5. B, D, F, H, J and L are light field photomicrographs. Other abbreviations used are c: cartilage primordium, g: gut, cp: choroid plexus.

given that these two growth factors have the same receptor specificity. Interestingly, VEGF-D does not induce vascular permeability [31].

VEGFR-3

VEGFR-3 is a receptor tyrosine kinase thought to signal for lymphangiogenesis, based on its expression on lymphatic endothelial cells. The extracellular domain of VEGFR-3 consists of seven immunoglobulin-homology domains and exhibits 37% amino acid sequence identity with VEGFR-2, whereas the tyrosine kinase domains of these receptors are 80% identical [32]. Two VEGFR-3 transcripts have been described – one of 5.8 kb in which the 3′coding region is 65 amino acids longer and encodes the long form of VEGFR-3 and one of 4.5 kb encoding the short form [33]. Only the longer form of VEGFR-3 is detected in human cell lines [33] and in mouse tissues [34]. VEGFR-3 undergoes post-translational processing, involving glycosylation and proteolysis, to yield a mature receptor composed of two polypeptides linked by disulfide bonds [35].

In mouse embryos, VEGFR-3 mRNA is localized at E8.5 in the developing blood vessels and in the angioblasts of the head mesenchyme. During days 11.5-12.5 of development, the VEGFR-3 transcript is more prominent in the developing veins and presumptive lymphatic endothelia than in arterial endothelia. VEGFR-3 transcripts become confined to the lymphatic endothelium during day 14.5 of development [27]. The proximity of VEGFR-3 positive endothelia and regions positive for VEGF-C and VEGF-D transcripts at various sites in the developing mouse embryo suggests a paracrine relationship between these growth factors and VEGFR-3 for control of lymphatic development.

Recent studies demonstrated expression of VEGFR-3 in human tumors. VEGFR-3 mRNA was abundant in the lymphatic sinuses of metastatic lymph nodes and in lymphangioma [27]. VEGFR-3 protein was found in the endothelium of lymphatic vessels around lymphomas. In cutaneous nodular AIDS-associated Kaposi's sarcoma, the tumor cells and the endothelium around the nodules were VEGFR-3 positive [36]. VEGFR-3 is expressed in lymphatic endothelia both in normal breast and in breast carcinomas, but it is also weakly expressed in the blood capillary endothelium, particularly in angiogenic capillaries [37]. VEGFR-3 is expressed specifically in lymphatic vessels in normal tissues but in benign and malignant vascular tumors VEGFR-3 is not a specific marker for lymphatic endothelium as it is up-regulated on neoplastic blood vessels [23].

VEGF-C and VEGF-D induce autophosphorylation of VEGFR-3 [11,24]. In the long form of VEGFR-3, tyrosine 1337 is a potential autophosphorylation site [38], and is a docking site for GRB2 and SHC involved in VEGFR-3 signal transduction [39]. In the HEL cell line expressing high levels of VEGFR-3, VEGF-C stimulation resulted in phosphorylation and activation of the recently identified "related adhesion focal tyrosine kinase" (RAFTK) and enhanced association of this kinase with the adaptor protein GRB2 [40]. The tyrosine phosphorylated molecules SHC, GRB2, and SOS have been found to form a signaling complex with the activated VEGFR-3 [40]. Upon stimulation of VEGFR-3 expressing cells, VEGF-C and VEGF-D induce rapid tyrosine phosphorylation of the SHC protein and activation of the mitogen-activated protein kinases ERK1 and ERK2 [14]. VEGFR-3 has been implicated in signaling mitogenesis in some endothelial cells *in vitro* as VEGF-C induces proliferation of porcine aortic endothelial cells expressing VEGFR-3 but not VEGFR-2 or VEGFR-1 [22]. Cell shape changes, actin reorganization and chemotaxis were observed upon stimulation with VEGF-C in cells overexpressing VEGFR-3 [22], but it appears that activation of VEGFR-3 is not sufficient to mediate vascular permeability as a VEGF-C mutant, which binds to and activates VEGFR-3 but not VEGFR-2, does not induce vascular permeability *in vivo* [16].

During embryogenesis, VEGFR-3 has an essential role in the development of the cardiovascular system before the emergence of the lymphatic vessels. This is consistent with the observation that, during embryogenesis, VEGFR-3 is initially expressed on a broad range of endothelia and is subsequently enriched in the venous endothelium and then in the lymphatic endothelium. In VEGFR-3-null embryos there are no major defects in the differentiation of endothelial cells and in the processes of vasculogenesis and angiogenesis. However, vascular remodeling and maturation are abnormal in large vessels which have defective lumens. Fluid accumulation occurs in the pericardial cavity, and cardiovascular failure at E9.5 [41]. Also, the yolk sac vasculature is underdeveloped, lacking major blood vessels because of a failure of remodeling of the capillary network into complex vitelline vessels [41]. Analysis of the role of VEGFR-3 in the emergence of the lymphatics may require conditional knockout mice in which the *VEGFR-3* gene can be inactivated just prior to the appearance of the lymphatic vasculature.

Concluding Remarks

Over the past 10 years, much has been learnt about the molecular control of angiogenesis [42], but only recently have the first regulators of lymphangiogenesis been identified. The availability of VEGF-C and VEGF-D offers the opportunity to induce lymphangiogenesis in the clinic, which may be useful for treatment of lymphoedema. The expression of VEGF-C and VEGF-D in tumors ([43], Achen *et al.*, unpublished) raises the possibility of tumor lymphangiogenesis. Despite involvement of the lymphatics in tumor metastasis, little is known about the relationship between tumor cells and the lymphatic endothelium. The route by which a tumor metastasizes may, in part, be determined by the angiogenic/lymphangiogenic growth factors secreted by tumor cells that modulate the prevalence of vessels in a tumor. Specific inhibitors of VEGF-C, VEGF-D or VEGFR-3 will be required to address this important issue.

References

1. Leak LV (1970) Electron microscopic observations on lymphatic capillaries and the structural components of the connective tissue-lymph interface. Microvasc Res 2:361-391
2. Casley-Smith JR (1980) The fine structure and functioning of tissue channels and lymphatics. Lymphology 12:177-183
3. Cann SA, van Netten JP, Ashby TL, Ashwood-Smith MJ, van der Westhuizen NG (1995) Role of lymphagenesis in neovascularisation. Lancet 346:903
4. Witte MH, Way DL, Witte CL, Bernas M (1997) Lymphangiogenesis: mechanisms, significance and clinical implications. In: Goldberg ID, Rosen EM (eds) Regulation of angiogenesis. Birkhäuser Verlag, Basel: 65-112
5. Clark E, Clark E (1937) Observations on living mammalian lymphatic capillaries, their relation to the blood vessels. Am J Anat 60:253-298
6. Wilting J, Kurz H, Oh S-J, Christ B (1999) Angiogenesis and lymphangiogenesis: analagous mechanisms and homologous growth factors. In: Little CD, Mironov V, Sage EH (eds) Vascular morphogenesis: in vivo, in vitro, in mente. Birkhäuser Verlag, Basel
7. Reichert F (1926) The regeneration of the lymphatics. Arch Surg 13:871-881
8. Achen MG, Stacker SA (1998) The vascular endothelial growth factor family; proteins which guide the development of the vasculature. Int J Exp Path 79:255-265
9. McDonald NQ, Hendrickson WA (1993) A structural superfamily of growth factors containing a cystine knot motif. Cell 73:421-424
10. Oh S-J, Jeltsch MM, Birkenhäger R, McCarthy JEG, Weich HA, Christ B, Alitalo K, Wilting J (1997) VEGF and VEGF-C: specific induction of angiogenesis and lymphangiogenesis in the differentiated avian chorioallantoic membrane. Dev Biol 188:96-109
11. Joukov V, Pajusola K, Kaipainen A, Chilov D, Lahtinen I, Kukk E, Saksela O, Kalkkinen N, Alitalo K (1996) A novel vascular endothelial growth factor, VEGF-C, is a ligand for the Flt-4 (VEGFR-3) and KDR (VEGFR-2) receptor tyrosine kinases. EMBO J 15:290-298
12. Joukov V, Sorsa T, Kumar V, Jeltsch M, Claesson-Welsh L, Cao Y, Saksela O, Kalkkinen N, Alitalo K (1997) Proteolytic processing regulates receptor specificity and activity of VEGF-C. EMBO J 16:3898-3911
13. Lymboussaki A, Partanen TA, Olofsson B, Thomas-Crusells J, Fletcher CD, de Waal RM, Kaipainen A, Alitalo K (1998) Expression of the vascular endothelial growth factor C receptor VEGFR-3 in lymphatic endothelium in skin and vascular tumors. Am J Pathol 153:395-403
14. Taipale J, Makinen T, Arighi E, Kukk E, Karkkainen M, Alitalo K (1999) Vascular endothelial growth factor receptor-3. Curr Top Microbiol Immunol 237:85-96

15. Millauer B, Wizigmann-Voos S, Schnürch H, Martinez R, Moller NPH, Risau W, Ullrich A (1993) High affinity VEGF binding and developmental expression suggest Flk-1 as a major regulator of vasculogenesis and angiogenesis. Cell 72:835-846
16. Joukov V, Kumar V, Sorsa T, Arighi E, Weich H, Saksela O, Alitalo K (1998) A recombinant mutant vascular endothelial growth factor-C that has lost VEGFR-2 binding, activation, and vascular permeability activities. J Biol Chem 273:6599-6602
17. Kukk E, Lymboussaki A, Taira S, Kaipainen A, Jeltsch M, Joukov V, Alitalo K (1996) VEGF-C receptor binding and pattern of expression with VEGFR-3 suggests a role in lymphatic vascular development. Development 122:3829-3837
18. Sabin F (1909) The lymphatic system in human embryos, with a consideration of the morphology of the system as a whole. Am J Anat 9:43-91
19. Jeltsch M, Kaipainen A, Joukov V, Meng X, Lakso M, Rauvala H, Swartz M, Fukumura D, Jain RK, Alitalo K (1997) Hyperplasia of lymphatic vessels in VEGF-C transgenic mice. Science 276:1423-1425
20. Lee J, Gray A, Yuan J, Luoh SM, Avraham H, Wood WI (1996) Vascular endothelial growth factor-related protein: a ligand and specific activator of the tyrosine kinase receptor Flt4. Proc Natl Acad Sci USA 93:1988-1992
21. Witzenbichler B, Asahara T, Murohara T, Silver M, Spyridopoulos I, Magner M, Principe N, Kearney M, Hu J-S, Isner JM (1998) Vascular endothelial growth factor-C (VEGF-C/VEGF-2) promotes angiogenesis in the setting of tissue ischemia. Am J Pathol 153:381-394
22. Cao Y, Linden P, Farnebo J, Cao R, Eriksson A, Kumar V, Qi JH, Claesson-Welsh L, Alitalo K (1998) VEGF-C induces angiogenesis in vivo. Proc Natl Acad Sci USA 95:14389-14394
23. Partanen TA, Alitalo K, Miettinen M (1999) Lack of lymphatic vascular specificity of vascular endothelial growth factor receptor 3 in 185 vascular tumors. Cancer 86:2406-2412
24. Achen MG, Jeltsch M, Kukk E, Mäkinen T, Vitali A, Wilks AF, Alitalo K, Stacker SA (1998) Vascular endothelial growth factor D (VEGF-D) is a ligand for the tyrosine kinases VEGF receptor 2 (Flk-1) and VEGF receptor 3 (Flt-4). Proc Natl Acad Sci USA 95:548-553
25. Stacker SA, Stenvers K, Ceasar C, Vitali A, Domagala T, Nice E, Roufail S, Simpson RJ, Moritz R, Karpanen T, Alitalo K, Achen MG (1999) Biosynthesis of VEGF-D involves proteolytic processing which generates non-covalent dimers. J Biol Chem 274:32127-32136
26. Avantaggiato V, Orlandini M, Acampora D, Oliviero S, Simeone A (1998) Embryonic expression pattern of the murine *figf* gene, a growth factor belonging to platelet-derived growth factor/vascular endothelial growth factor family. Mech Dev 73:221-224
27. Kaipainen A, Korhonen J, Mustonen T, van Hinsbergh VW, Fang GH, Dumont D, Breitman M, Alitalo K (1995) Expression of the fms-like tyrosine kinase 4 gene becomes restricted to lymphatic endothelium during development. Proc Natl Acad Sci USA 92:3566-3570
28. Orlandini M, Marconcini L, Ferruzzi R, Oliviero S (1996) Identification of a c-*fos*-induced gene that is related to the platelet-derived growth factor/vascular endothelial growth factor family. Proc Natl Acad Sci USA 93:11675-11680
29. Saez E, Rutberg SE, Mueller E, Oppenheim H, Smoluk J, Yuspa SH, Spiegelman BM (1995) c-*fos* is required for malignant progression of skin tumors. Cell 82:721-732
30. Marconcini L, Marchio S, Morbidelli L, Cartocci E, Albini A, Ziche M, Bussolino F, Oliviero S (1999) *c-fos*-induced growth factor/vascular endothelial growth factor D induces angiogenesis *in vivo* and *in vitro*. Proc Natl Acad Sci USA 96:9671-9676
31. Stacker SA, Vitali A, Caesar C, Domagala T, Groenen LC, Nice E, Achen MG, Wilks AF (1999) A mutant form of vascular endothelial growth factor (VEGF) that lacks VEGF receptor-2 activation retains the ability to induce vascular permeability. J Biol Chem 274:34884-34892
32. Pajusola K, Aprelikova O, Korhonen J, Kaipainen A, Pertovaara L, Alitalo R, Alitalo K (1992) FLT4 receptor tyrosine kinase contains seven immunoglobulin-like loops and is expressed in multiple human tissues and cell lines. Cancer Res 52:5738-5743
33. Pajusola K, Aprelikova O, Armstrong E, Morris S, Alitalo K (1993) Two human FLT4 receptor tyrosine kinase isoforms with distinct carboxy terminal tails are produced by alternative processing of primary transcripts. Oncogene 8:2931-2937
34. Galland F, Karamysheva A, Pebusque M-J, Borg J-P, Rottapel R, Dubreuil P, Rosnet O,

Birnbaum D (1993) The *FLT4* gene encodes a transmembrane tyrosine kinase related to the vascular endothelial growth factor receptor. Oncogene 8:1233-1240

35. Borg JP, deLapeyriere O, Noguchi T, Rottapel R, Dubreuil P, Birnbaum D (1995) Biochemical characterization of two isoforms of FLT4, a VEGF receptor-related tyrosine kinase. Oncogene 10:973-984
36. Jussila L, Valtola R, Partanen TA, Salven P, Heikkilä P, Matikainen MT, Renkonen R, Kaipainen A, Detmar M, Tschachler E, Alitalo R, Alitalo K (1998) Lymphatic endothelium and Kaposi's sarcoma cells detected by antibodies against VEGFR-3. Cancer Res 58:1599-1604
37. Valtola R, Salven P, Heikkila P, Taipale J, Joensuu H, Rehn M, Pihlajaniemi T, Weich H, deWaal R, Alitalo K (1999) VEGFR-3 and its ligand VEGF-C are associated with angiogenesis in breast cancer. Am J Pathol 154:1381-1390
38. Fournier E, Rosnet O, Marchetto S, Turck CW, Rottapel R, Pelicci PG, Birnbaum D, Borg J-P (1996) Interaction with the phosphotyrosine binding domain/phosphotyrosine interacting domain of SHC is required for the transforming activity of the FLT4/VEGFR3 receptor tyrosine kinase. J Biol Chem 271:12956-12963
39. Fournier E, Dubreuil P, Birnbaum D, Borg J-P (1995) Mutation at tyrosine residue 1337 abrogates ligand-dependent transforming capacity of the FLT4 receptor. Oncogene 11:921-931
40. Wang J-F, Ganju RK, Liu Z-Y, Avraham H, Avraham S, Groopman JE (1997) Signal transduction in human hematopoietic cells by vascular endothelial growth factor related protein, a novel ligand for the FLT4 receptor. Blood 90:3507-3515
41. Dumont DJ, Jussila L, Taipale J, Lymboussaki A, Mustonen T, Pajusola K, Breitman M, Alitalo K (1998) Cardiovascular failure in mouse embryos deficient in VEGF receptor-3. Science 282:946-949
42. Risau W (1997) Mechanisms of angiogenesis. Nature 386:671-674
43. Salven P, Lymboussaki A, Heikkilä P, Jääskela-Saari H, Enholm B, Aase K, von Euler G, Eriksson U, Alitalo K, Joensuu H (1998) Vascular endothelial growth factors VEGF-B and VEGF-C are expressed in human tumors. Am J Pathol 153:103-108
44. Partanen TA, Makinen T, Arola J, Suda T, Weich HA, Alitalo K (1999) Endothelial growth factor receptors in human fetal heart. Circulation 100:583-586

Hemangioblastic Precursors in the Avian Embryo

A. Eichmann, C. Corbel, L. Pardanaud, C. Bréant, D. Moyon and L. Yuan
Institut d'Embryologie Cellulaire et Moleculaire CNRS FRE 2160
49bis, Avenue de la Belle Gabrielle, 94736 Nogent-sur-Marne Cedex, France

Introduction

Endothelial and hemopoietic cells of the avian embryonic yolk sac arise from hemangioblastic aggregates, which are clusters of mesodermal cells. These aggregates are present at the border of the embryonic and extraembryonic area from the beginning of the second day of the incubation period (around the 1-somite stage). They subsequently mature to blood islands, the external cells of which flatten and differentiate into endothelial cells (EC), while the internal cells of the cluster form hemopoietic cells (HC). Evidence for blood island formation was initially obtained by observations of living chick blastoderms (Sabin 1920; Murray 1932) (Fig. 1). Based on the simultaneous emergence of EC and HC in the blood islands, it was proposed that the two cell types originate from a common precursor, termed hemangioblast. Recent evidence, which will be summarized here, suggests that hemangioblastic precursors can be identified based on their expression of vascular endothelial growth factor receptor 2 (VEGFR2).

VEGFs and their Receptors

VEGFR2 is one of three endothelial-specific receptor tyrosine kinases, VEGFR1-3, all characterized by the presence of seven extracellular immunoglobulin-like domains, a transmembrane domain and a split intracellular tyrosine kinase domain. The first identified ligand of this receptor family, VEGF, is currently considered as the major inducer of angiogenesis in both normal and pathological conditions, based on its EC-specific biological effects which include stimulation of proliferation, migration and tube formation as well as regulation of vascular permeability (for review see Neufeld et al. 1999). Targeted inactivation of the

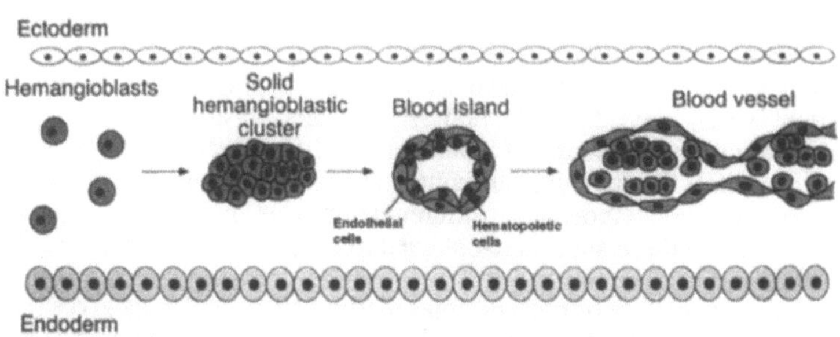

Fig. 1. Yolk sac blood island development according to Sabin (1920). Following gastrulation, mesodermal hemangioblastic precursors migrate to the yolk sac and give rise to solid hemangioblastic aggregates. The peripheral cells of the aggregates differentiate to EC and HC appear in the lumen. Blood island anastomosis marks the beginning of the embryonic circulation.

VEGF gene in the mouse has demonstrated the necessity of this growth factor for the development of the vascular system: inactivation of one allele of the VEGF gene is lethal between E9.5 and E10.5 due to insufficient development of the endothelial network, indicating that concentration of this factor is critical (Ferrara et al. 1996; Carmeliet et al. 1996).

Several growth factors closely related to VEGF have been identified, including placenta-growth factor (PlGF), VEGF-B, -C, and -D (for review see Korpelainen and Alitalo 1998). VEGFs generally act as paracrine growth factors, they are expressed by mesenchymal and epithelial cells adjacent to EC expressing their receptors. VEGF binds to VEGFR1 and 2, while PlGF and VEGF-B bind only VEGFR1. VEGF-C and -D bind to VEGFR2 and 3. We have cloned the equivalents of VEGF-C, VEGFR2 and VEGFR3 in the quail (Eichmann et al. 1996, 1998). Binding studies suggest that the receptor-ligand specificities of these molecules are conserved in birds (Eichmann et al. 1998).

Posterior Mesodermal VEGFR2 Positive Cells have Hemangioblastic Potential

Among the VEGF receptors, VEGFR2 appears as the earliest endothelial lineage marker, as determined by *in situ* hybridization studies on mouse and avian embryos (Yamaguchi et al. 1993; Millauer et al. 1993; Eichmann et al. 1993). Moreover, targeted gene inactivation experiments have shown that the function of VEGFR2 is crucial for the emergence of both EC and HC in the embryo, since both cell types are absent in homozygous null mutant mice (Shalaby et al, 1995). Murray (1932) mapped the 'hemangioblast' potential of the primitive streak-stage avian embryo to the posterior two thirds of the embryo. This corresponds to the territory which displays strong VEGFR2 expression (Fig. 2A), suggesting that VEGFR2 expressing cells might be hemangioblasts.

To address this question, cell suspensions from posterior territories of chick embryos at the gastrulation stage were labeled with a monoclonal antibody directed against the extracellular domain of VEGFR2. VEGFR2$^+$ cells were then sorted by FACS and cultured clonally in semi-solid medium (Eichmann et al. 1997). In the absence of VEGF, VEGFR2$^+$ cells gave rise to HC colonies at a frequency of one HC colony for 10 precursors seeded. Different hemopoietic lineages could be obtained: erythrocytes as well as thrombocytes formed in the basal conditions (10% bovine plasma, no growth factors added; Cormier and Dieterlen-Lièvre 1988), whereas macrophages differentiated in GM-CSF-supplemented medium. In the presence of VEGF, EC differentiation was induced in a dose-dependent manner, concomitant with a reduction of hemopoietic differentiation. At a VEGF dose inducing maximal endothelial differentiation (one endothelial colony for 10 precursors seeded), hemopoietic differentiation was reduced to about 50% of control values. Similar results were obtained in the presence of VEGF-C (Eichmann et al. 1998).

Thus, the population of posterior mesodermal VEGFR2$^+$ cells gives rise to both EC and HC and can therefore be considered as hemangioblastic cells. One VEGF-R2$^+$ cell can, however, give rise to either one or the other, but not to both cell types. Endothelial differentiation depends on the presence of VEGF, whereas the hemopoietic phenotype appears in the absence of VEGF. The VEGFR2 knockout phenotype in mice had, however, suggested that hemopoietic differentiation also depended on receptor activity (Shalaby et al. 1995). Indeed, when VEGFR2$^+$ cells

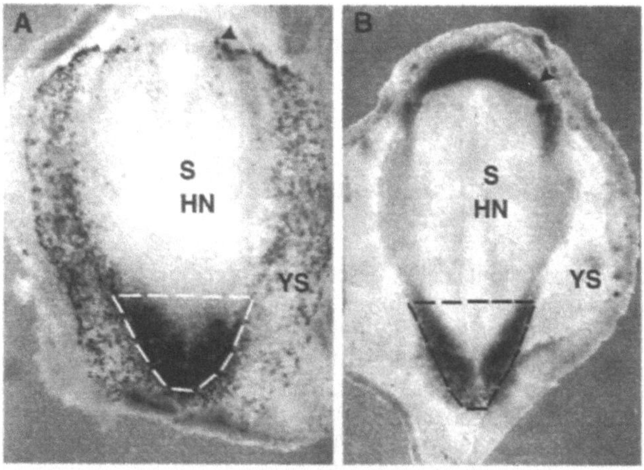

Fig. 2 **A**. VEGFR2 expression in hemangioblastic precursors. 1-somite stage chick embryo hybridized *in toto* with VEGFR2 antisense riboprobe. Note expression in posterior region of the embryo (stippled area), which was used for culture experiments and subtractive hybridization.
B. Expression of sizzled in a 1-somite stage embryo. Sizzled and VEGFR2 are co-expressed in posterior mesodermal hemangioblastic cells. Outside this region, no co-expression is observed: sizzled is expressed in the procardiac region (arrowhead) and absent from embryonic or yolk sac VEGFR2⁺ angioblasts. HN: Hensen's node, s: somite, YS: yolk sac.

were cultured in the presence of soluble extracellular domain of VEGFR2, hemopoietic differentiation was reduced in a dose-dependent manner to about 20% of control values (Eichmann et al. 1997). These data are consistent with the presence of an unidentified VEGFR2 ligand in the medium, which would be responsible for hemopoietic differentiation and titrated out by increasing receptor concentrations.

These experiments are equally consistent with two hypotheses: either committed precursors for the two lineages exist, which differ in their expression of other surface molecules, or the population may be homogenous and the hemangioblasts are faced with a developmental choice, to enter the endothelial or the hemopoietic pathway.

A common precursor for both cell types has recently been described in cultures from mouse ES cells (Choi et al. 1998; Nishikawa et al. 1998a). In one study, ES cells carrying two different lineage markers were mixed together and allowed to develop to blast colonies with hemopoietic potential. Each blast colony was derived from either one of the two types of ES cells, suggesting a clonal origin (Kennedy et al. 1997). Subsequently, it was shown that a single blast cell colony can also give rise to EC, when placed in the appropriate culture conditions (Choi et al. 1998). In the other study, ES cells were sorted using a monoclonal antibody directed against mouse VEGFR2 and cultured in medium containing hemopoietic cytokines on a stromal cell layer of OP9 cells. In these conditions, colonies composed of EC and HC developed from a single VEGFR2⁺ cell (Nishikawa et al. 1998a). Two recent studies using VEGFR2 deficient ES cells show that in certain *in vitro* conditions, VEGFR2 deficient cells are able to differentiate into EC and HC, suggesting a role of VEGFR2 in the proliferation/migration of hemangioblastic precursors (Schuh et al. 1999; Hidaka et al. 1999). The precise role of VEGFR2 in hemangioblast development remains to be defined. In addition, it remains to be determined if

VEGFR2^{+} cells may have other developmental potentialities as well, which could not be expressed in the *in vitro* systems used to analyse avian precursors or murine ES cells.

EC and HC Differentiation at Later Embryonic Stages

Following the gastrulation stages described above, the embryonic mesoderm is separated into two layers by the formation of the coelomic cavity, the superficial somatopleural layer in contact with the ectoderm and the deep splanchnopleural layer in contact with the endoderm. As shown in avian chimeras, the splanchnopleural mesoderm has the capacity to give rise to both EC and HC (Pardanaud and Dieterlen-Lièvre, 1993). In contrast, the somatopleural mesoderm has no potential to produce EC. The EC which vascularize somatopleural derivatives such as the body wall and the limbs derive from the somites. Somites, however, do not give rise to HC (Pardanaud et al. 1996) (Fig. 3). The factor(s) responsible for the difference in developmental potentialities of the two mesodermal layers might be produced by the endoderm and the ectoderm. If the somatopleural mesoderm is co-cultured with endoderm for 24 hours, it acquires the capacity to give rise to EC and HC. Conversely, when splanchnopleural mesoderm is associated with the ectoderm, its capacity to give rise to HC is abolished and the production of EC is greatly reduced (Pardanaud and Dieterlen-Lièvre, 1999). Several growth factors, including bFGF and VEGF, can mimic the effect of the endoderm, while others, such as EGF or TGFα mimic the effect of the ectoderm (Pardanaud and Dieterlen-Lièvre, 1999).

The splanchnopleural precursors migrate to the ventral aspect of the aorta, where they produce HC in the intraaortic clusters. In the avian, these clusters are present between the third and fourth day of development (E3-E4) and in histological stainings they appear to bud from the ventral aortic endothelium (Dieterlen-Lièvre and Martin 1981). Jaffredo et al. (1998) have performed intra-cardiac injections of Di-Ac-LDL, which is taken up by EC of the aortic wall. Double-labeling of

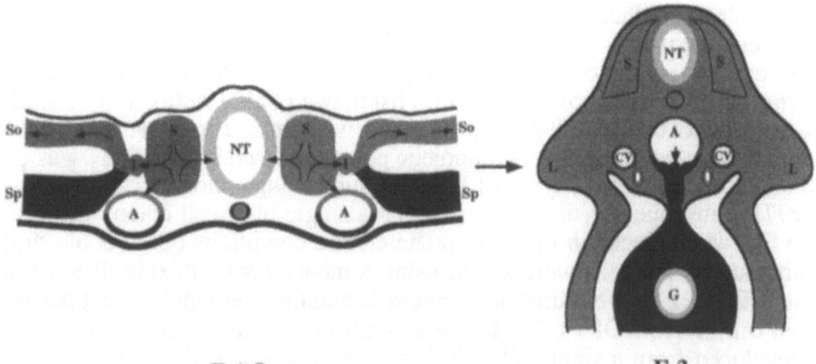

Fig. 3. The capacity of mesodermal cells to give rise to HC and EC following coelome formation is restricted to the splanchnopleural mesoderm (Sp). Somitic mesoderm (S) gives rise to EC only. The aorta is of mosaic origin, somitic for its dorsal aspect, and splanchnopleural for its ventral aspect. A: aorta; CV: cardinal vein; G: gut; I: intermediate mesoderm; L: limb; NT: neural tube; So: somatopleural mesoderm.

embryos with anti-VEGFR2 and CD45, the pan-leukocytic surface antigen, showed that only EC were present in the floor of the aorta at the time of injection (E2). One day later, Di-Ac-LDL positive cells had, however, formed CD45 positive HC clusters, suggesting an endothelial origin of these cells. In the mouse embryo, Nishikawa et al. (1998b) have reported the generation of lymphohemopoietic cells from EC sorted out from E9.5 mouse embryos or yolk sacs.

Caprioli et al. (1998) have examined the developmental potential of the allantois, one of the embryonic appendages of amniotes. The allantoic bud was retrieved from a quail embryo prior to vascularization and grafted into the coelom of a chick host. The graft emitted HC as well as EC progenitors which both seeded the host bone marrow. It remains to be determined if the migrating cells are bipotent hemangioblasts or monopotent progenitors of each lineage.

In the adult, VEGFR2$^+$ cells constitute a small fraction of bone-marrow HC with stem cell potential (Ziegler et al. 1999). These cells furthermore have the capacity to differentiate to EC (for review see Rafii, 2000).

Conclusions and Perspectives

The experiments described above suggest that expression of VEGFR2, an otherwise EC-restricted molecule, immediately precedes HC formation, both in the early gastrulating embryo as well as in the intraaortic clusters and perhaps even in the adult. However, several questions remain to be answered: is the VEGFR2$^+$ population homogenous and what are the mechanisms triggering hemopoietic differentiation of these precursors? It is possible that one or several different strategies may be employed to generate HC in the early yolk sac and in the embryo proper.

To address the first two questions, we have begun a search for genes expressed specifically by VEGFR2$^+$ cells.

Molecular Cloning of VEGFR2$^+$ Specific Genes

Cell-sorting of VEGFR2-labeled cells from 60 posterior primitive streaks yields around $2x10^4$ positive and negative cells, with a purity of about 95% (Eichmann et al. 1997, 1998). After extraction of the RNA from these cells, cDNA was synthetized using a PCR-based technique and subtractive hybridization was applied to isolate genes specific for the VEGFR2$^+$ cell population. The subtracted cDNA from the positive cells was cloned into a plasmid vector and subjected to a differential screening with subtracted probes (L. Yuan, C. Bréant and A. Eichmann, unpublished results).

Cross-hybridization of individual clones led to the identification of one major clone, representing close to 30% of differentially expressed cDNAs. The full-size cDNA of this clone was obtained by screening nonsubtracted cDNA. Comparison of the deduced amino-acid sequence showed 66% amino-acid identity to the *sizzled* gene, recently cloned in *Xenopus laevis* (Salic et al. 1997). No other frequently represented clone was found.

Sizzled is a member of receptors for growth factors of the *wnt* family, which constitute a large family of signaling molecules (20 members identified) playing important roles during embryonic development as well as in tumor growth (for review see Cadigan and Nusse 1997). *Wnt* actions are mediated by binding to seven-transmembrane receptors of the *frizzled* family (Wang et al. 1996). The *wnt*-binding domain of *frizzled* receptors is rich in cysteine residues (CRD: Cysteine Rich Domain). *Sizzled* contains this CRD, but no transmembrane domain, suggesting a potential role as a *wnt* antagonist. In *Xenopus* embryos, *sizzled* is indeed capable of antagonizing Xwnt8 actions (Salic et al. 1997).

In situ hybridization with chick *sizzled* riboprobes to gastrulation-stage embryos showed that *sizzled* is co-expressed with VEGFR2 in posterior mesodermal cells (Fig. 2B). Moreover, expression of this gene is restricted to the hemangioblastic precursors, neither EC nor HC differentiating in the yolk sac blood islands are *sizzled* positive (D. Moyon and A. Eichmann, unpublished results).

The remaining cDNA clones are currently being analysed by sequencing and *in situ* hybridization. Among several dozen clones sequenced to date, most are unknown, while some are homologues of genes already identified in different species. Among those, we retained for further analysis the *septin-2* gene. Septins were first identified in yeast (*S. cerevisiae*), but are conserved throughout evolution and have been cloned in Drosophila and mammals (for review see Longtine et al. 1996). Septin proteins show GTPase activity and form 10 nm intermediate filaments which are localized at the cleavage plane of dividing cells. In budding yeast, septins are involved in asymmetric cell divisions, as shown by loss-of-function mutants. In Drosophila, loss of function of the septin 'peanut' leads to multi-nuclearity in imaginal disc cells, suggesting that septin function in cell division has been conserved in evolution and is not restricted to asymmetric cell divisions. The *septin* clone we isolated in the chick is most homologous to human *septin-2* (L. Yuan, D. Moyon and A. Eichmann, unpublished results). *Septin-2* is expressed by VEGFR2$^+$ cells of the posterior primitive streak, as shown by *in situ* hybridization. At later stages, *septin-2* is maintained in some EC and HC of the developing yolk sac blood islands, as well as in intra-aortic hemopoietic clusters, suggesting a possible function of this gene in the division of these cells.

A zinc-finger transcription factor, *Vezf1* (vascular endothelial zinc finger 1), expressed exclusively in EC and their precursors has recently been described in the mouse (Xiong et al. 1999). In order to test whether this transcription factor was present in VEGFR2$^+$ mesodermal precursors, we designed oligonucleotides corresponding to the murine sequence and performed PCR on VEGFR2+ and - cell subtracted cDNA. A fragment of 650 bp was amplified from positive cell cDNA, but not from negative cell cDNA. Sequence analysis of this fragment showed high homology to murine *Vezf1* (L. Yuan and A. Eichmann, unpublished results). Expression analysis of this gene is currently underway.

Functional analysis of the isolated genes will be carried out in the avian embryo using gene transfer by retroviral strategies or electroporation (Sakamoto et al. 1998).

References

Cadigan KM, Nusse R (1997) Wnt signaling: a common theme in animal development. Genes Dev 11:3286-3305

Caprioli A, Jaffredo T, Gautier R, Dubourg C, Dieterlen-Lièvre F (1998) Blood-borne seeding by hematopoietic and endothelial precursors from the allantois. Proc Natl Acad Sci USA 95:1641-1646

Carmeliet P, Ferreira V, Breier G, Pollefeyt S, Kieckens L, Gertsenstein M, Fahrig M, Vandenhoeck A, Harpal K, Eberhardt C, Declercq C, Pawling J, Moons L, Collen D, Risau W, Nagy A (1996) Abnormal blood vessel development and lethality in embryos lacking a single VEGF allele. Nature 380:435-439

Choi K, Kennedy M, Kazarov A, Papadimitriou JC, Keller G (1998) A common precursor for hematopoietic and endothelial cells. Development 125:725-732

Cormier F, Dieterlen-Lièvre F (1988) The wall of the chick embryo aorta harbours M-CFC, G-CFC, GM-CFC and BFU-E. Development 102:279-285

Dieterlen-Lièvre F, Martin C (1981) Diffuse intraembryonic hemopoiesis in normal and chimeric avian development. Dev Biol 88:180-191

Eichmann A, Marcelle C, Breant C, Le Douarin NM (1993) 2 molecules related to the VEGF receptor are expressed in early endothelial cells during avian embryonic development. Mech Dev 42:33-48

Eichmann A, Marcelle C, Breant C, Le Douarin, NM (1996) Molecular cloning of Quek 1 and 2, two avian vascular endothelial growth factor (VEGF) receptor-like molecules. Gene 174:3-8

Eichmann A, Corbel C, Nataf V, Vaigot P, Bréant C, Le Douarin NM (1997) Ligand-dependent development of the endothelial and hemopoietic lineages from embryonic mesodermal cells expressing vascular endothelial growth factor receptor 2. Proc Natl Acad Sci USA 94:5141-5146

Eichmann A, Corbel C, Jaffredo T, Bréant C, Joukov V, Kumar V, Alitalo K, le Douarin NM (1998) Avian VEGF-C: cloning, embryonic expression pattern and stimulation of the differentiation of VEGFR2-expressing endothelial cell precursors. Development 125:743-752

Ferrara, N, Carver-Moore K, Chen H, Dowd M, Lu L, O'Shea KS, Powell-Braxton L, Hillan KJ, Moore MW (1996) Heterozygous embryonic lethality induced by targeted inactivation of the VEGF gene. Nature 380:439-442

Hidaka M, Stanford WL, Bernstein A (1999) Conditional requirement for the Flk-1 receptor in the in vitro generation of early hematopoietic cells. Proc Natl Acad Sci USA 96:7370-7375

Jaffredo T, Gauthier R, Eichmann A, Dieterlen-Lièvre F (1998) Intraaortic hemopoietic cells are derived from endothelial cells during ontogeny. Development 125:4575-4583

Kennedy M, Firpo M, Choi K, Wall C, Robertson S, Kabrun N, Keller G (1997) A common precursor for primitive erythropoiesis and definitive haematopoiesis. Nature 386:488-493

Korpelainen EI, Alitalo K (1998) Signaling angiogenesis and lymphangiogenesis. Current Op Cell Biol 10:159-164

Longtine MS, DeMarini DJ, Valencik ML, Al-Awar OS, Fares H, De Virgilio C, Pringle JR (1996) The septins: roles in cytokinesis and other processes. Curr Op Cell Biol 8:106-119

Millauer B, Wizigmann-Voos S, Schnürch H, Martinez R, Moller NP, Risau W, Ullrich A (1993) High affinity VEGF binding and developmental expression suggest Flk-1 as a major regulator of vasculogenesis and angiogenesis. Cell 72:835-846

Murray PDF (1932) The development in vitro of blood of the early chick embryo. Proc Roy Soc B Vol III:497-521

Neufeld G, Cohen T, Gengrinovitch S, Poltorak Z (1999) Vascular endothelial growth factor (VEGF) and its receptors. FASEB J 13:9-22

Nishikawa SI, Nishikawa S, Hirashima M, Matsuyoshi N, Kodama H (1998a) Progressive lineage analysis by cell sorting and culture identifies FLK-1+VE-cadherin+ cells at a diverging point of endothelial and hemopoietic lineages. Development 125:1747-1757

Nishikawa SI, Nishikawa S, Kawamoto H, Yoshida H, Kizumoto M, Kataoka H, Katsura,Y (1998b) In vitro generation of lymphohematopoietic cells from endothelial cells purified from murine embryos. Immunity 8:761-769

Pardanaud L, Dieterlen-Lievre F (1993) Emergence of endothelial and hemopoietic cells in the avian embryo. Anat Embryol 187:107-114

Pardanaud L, Luton D, Prigent M, Bourcheix LM, Catala M, Dieterlen-Lièvre F (1996) Two distinct endothelial lineages in ontogeny, one of them related to hemopoiesis. Development 122:1363-1371

Pardanaud L, Dieterlen-Lièvre F (1999) Manipulation of the angiopoietic/hemangiopoietic commitment in the avian embryo. Development 126:617-627

Rafii S (2000) Circulating endothelial precursors: mystery, reality, and promise. J Clin Invest 105:17-19

Sabin, FR (1920) Studies on the origin of blood vessels and of red blood corpuscles as seen in the living blastoderm of chicks during the second day of incubation. Contrib Embryol Carnegie Institute Pub 9-36:214-262

Salic A, Kroll KL, Evans LM, Kirschner M (1997) Sizzled: a secreted Xwnt8 antagonist expressed in the ventral marginal zone of Xenopus embryos. Development 124:4739-4748

Sakamoto K, Nakamura H, Tagaki M, Takeda S, Katsube K (1998) Ectopic expression of lunatic fringe leads to downregulation of serrate-1 in the developing chick neural tube: analysis using in ovo electroporation transfection technique. FEBS Letters 426:337-341

Schuh AC, Faloon P, Hu QL, Bhimani M, Choi K (1999) In vitro hematopoietic and endothelial potential of flk-1-/- embryonic stem cells and embryos. Proc Natl Acad Sci USA 96:2159-2164

Shalaby F, Rossant J, Yamaguchi TP, Gertsenstein M, Wu XF, Breitman ML, Schuh AC (1995) Failure of blood island formation and vasculogenesis in Flk-1 deficient mice. Nature 376:62-66

Wang Y, Macke JP, Abella BS, Andreasson K, Worley P, Gilbert D, Copeland NG, Jenkins NA, Nathans J (1996) A large family of putative transmembrane receptors homologous to the product of the Drosophila tissue polarity gene frizzled. J Biol Chem 271:4468-4476

Yamaguchi TP, Dumont DJ, Conlon RA, Breitman ML, Rossant J (1993) flk-1, an flt-related receptor tyrosine kinase is an early marker for endothelial cell precursors. Development 118:489-498

Xiong J-W, Leahy A, Lee HH, Stuhlmann H (1999) Vezf1: A Zn finger transcription factor restricted to endothelial cells and their precursors. Dev Biol 206:123-141

Ziegler BL, Valtieri M, Porada GA, De Maria R, Muller R, Masella B, Gabbianelli M, Casella I, Pelosi E, Bock T, Zanjani ED, Peschle C (1999) KDR receptor: a key marker defining hematopoietic stem cells. Science 285:1553-1558

Cloning of JAM-2 and JAM-3: an Emerging Junctional Adhesion Molecular Family?

M.A. Aurrand-Lions[1], L. Duncan[1,2], L. Du Pasquier[3] and B. A. Imhof[1]

[1] Department of Pathology, Centre Medical Universitaire,1 rue Michel-Servet, CH-1211 Geneva, Switzerland

[2] Present address: Wellcome Trust/MRC building, Addenbrookes Hospital, Hills Rd, Cambridge CB2 2XY, England

[3] Basel Institute for Immunology, Grenzacherstrasse 487, CH-4005 Basel, Switzerland

Introduction

Throughout embryonic and early postnatal development, endothelial cells proliferate and differentiate to form new blood vessels via vasculogenesis and angiogenesis (1, 2). In adult organisms the endothelium defines the blood-tissue barrier and consists of non-cycling quiescent cells. These polarized cells are linked to each other by tight junctions and adherens junctions to form a continuous layer of cells. JAM (hereafter referred as JAM-1) and VE-cadherin were characterized as endothelial adhesion molecules participating in tight and adherens intercellular junctions respectively (3, 4). These molecules were shown to regulate vascular functions such as monocyte transmigration or paracellular permeability, probably as a consequence of their structural contribution to the vessel wall. It is well established that the regulated and coordinated expression of adhesion molecules is necessary to maintain normal vascular functions such as tissue homeostasis, vascular permeability, leukocyte emigration, fibrinolysis, coagulation and vasotonus. Temporary changes in the adhesion properties of vascular endothelium have been observed during inflammation, tumor growth, wounding or angiogenesis (5-7). The presence of a growing tumor increases the local concentration of angiogenic factors leading to a switch from non-cycling quiescent endothelial cells to proliferating endothelium. As a result, endothelial cells of existing vessels degrade the extracellular matrix (ECM) and invade the surrounding tissue, which leads to vascularization of tumors. During the angiogenic switch, the pattern of endothelial gene expression is modified, and the level of transcripts encoding proteins involved in cell migration or cell division is affected. For example, the balance between proteases/antiproteases such as PA/PAI1 is changed, leading to increased endothelial cell motility (8). Moreover, the treatment of endothelial cells with angiogenic factors results in a fourfold increase in $\alpha_V\beta_3$ integrin expression, an adhesion molecule implicated in cell migration (9). In addition, angiogenesis was shown to modify the endothelial inflammatory response leading to abnormal expression of inflammatory adhesion molecules (10, 11). These examples do not constitute an exhaustive list, but support the hypothesis that during the angiogenic switch, the global adhesion behavior of endothelial cells is changed.

Since many other adhesion molecules, such as immunoglobulin superfamily molecules (Ig Sf), may be regulated during this process, we began looking for new members of this family that might be regulated during angiogenesis. For this purpose we used an in vitro model based on the coculture of endothelial cell lines with melanoma cells, previously demonstrated to induce modifications of the endothelial cell phenotype (12). We developed a differential display technique that allowed the specific identification of transcripts encoding surface molecules of the Ig Sf. This was achieved by running RNA display with degenerated primers that were specific for a consensus sequence found in C_2 domains of Ig Sf molecules. Using this method, we identified a novel endothelial transcript encoding an Ig Sf molecule, downregulated in endothelial cells that reach confluency. Sequence comparisons demonstrated the recently cloned JAM-1 (3, 13) to be the closest related sequence, and we therefore named the new molecule JAM-2. Further sequence comparisons of JAM-1 and JAM-2 with EST databases identified another member of this new molecular family: JAM-3. Sequence alignments of JAM-1, JAM-2, and JAM-3 reveal a structural organization of the new molecular family in type I proteins comprising one V and one C_2 domain, reminiscent of the recently described CTX molecular family (14, 15). Specific and common properties of JAM-1, JAM-2 or JAM-3 will be discussed in the context of a novel protein family subset: the Junctional Adhesion Molecular family.

Material and methods

"Targeted differential display"

Adhesion molecules of the Ig Sf possess structural domains similar to the variable (V) or constant (C) immunoglobulin domains found in T cell and B cell receptors (16). The alignment of the C_2 domains of several adhesion molecules allowed the identification of a linear consensus, surrounding the cysteine residue participating in the C_2 domain structure (Fig 1A). Most of the C_2 domains are characterized by two cysteine residues and a single disulfide bridge, but some molecules (Titin, CD2, CTX) possess C_2 domains with two disulfide bridges and four cysteine residues (14). We used the most frequent sequences (YRCXAS, YQCXAS, YYCXAS) surrounding the cysteine residue conserved in all C_2 domains to design degenerated primers. Upon reverse translation of amino-acid sequences, the degeneracy of the primers was given by positions where purine (R), pyrimidin (Y) or any of the four bases (N) were placed (Fig 1B). These primers targeted the C_2 consensus sequence in forward orientation, and may prime in reverse orientation on a limited number of sequences because of their degeneracy. This resulted in discrete products when they were used for polymerase chain reaction (PCR). The primers were used in a differential PCR approach to amplify the endothelial cDNA encoding for adhesion molecule comprising C_2 domains, which may be regulated during the coculture of endothelial cells with tumor cells.

A

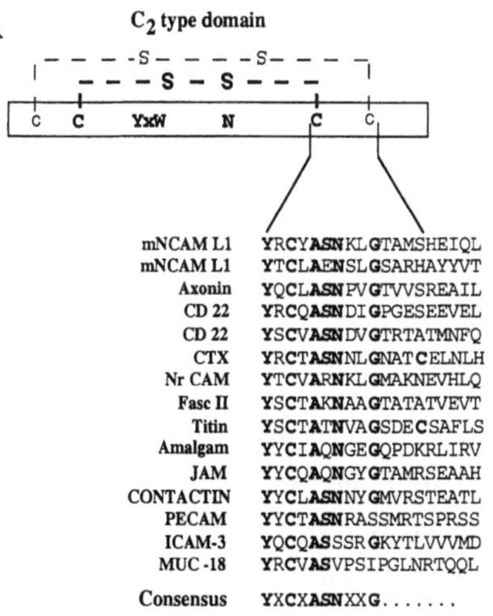

B

```
Y   R   C   X   A   S1
TAY AGN TGY NNN GCY TCY AA

Y   R   C   X   A   S2
TAY AGN TGY NNN GCY AGY AA

Y   Q   C   X   A   S1
TAY CRG TGY NNN GCY TCY AA

Y   Q   C   X   A   S2
TAY CRG TGY NNN GCY AGY AA

Y   Y   C   X   A   S1
TAY TAY TGY NNN GCY TCY AA

Y   Y   C   X   A   S2
TAY TAY TGY NNN GCY AGY AA
```

Fig. 1: A: Schematic representation of the C_2 like domain and alignment of protein sequences surrounding the canonical cysteine residues. B: Reverse translation of the most frequent amino-acid sequences. R is used for A or G (purine), Y for C or T (pyrimidine), N for any of the four bases.

Cloning and sequence analysis

Unknown sequences of PCR products obtained in the differential display were compared to murine EST database using tblastn. Consortium cDNA clones homologous to the PCR product sequences were obtained from the UK HGMP Resource Center in Cambridge and resequenced. Three EST corresponded to the sequence described in the present manuscript (Accession No. AA726206, AA052463 and AA175925), and one was identified as a related sequence (Acc N° AA445150). None of them was complete, and the remaining 5' coding sequences were obtained using the 5'RACE-PCR System for Rapid Amplification of cDNA Ends (Gibco BRL, Paisley, Scotland). We therefore named the transcript corresponding to the differential PCR product JAM-2, and the related sequence JAM-3. The cDNA encoding JAM-1 was kindly provided by P. Naquet (CIML, Marseille-Luminy, France). Sequence analysis and sequence comparisons were performed via the applications available on the Expasy Molecular Biology Server (Blast, Prosite, Swiss-Prot).

EGFP constructs

The cloning of JAM-2 and JAM-1 consisted in using the 3'HpaI or 3'ScaI respectively and the 5' EcoRI sites in order to clone the coding sequence for JAM-2 and JAM-1 in frame with the sequence encoding EGFP in the pcDNA3 vector. Since no equivalent restriction sites were available in the sequence of JAM-3, we used a PCR approach to obtain the JAM-3-EGFP chimeric molecule using Pfu DNA polymerase. All constructs were verified by sequencing on both strands.

Results and discussion

Identification of JAM-2 and JAM-3

The use of the degenerated primers (see Material and Method) in a differential display technique allowed the identification of one differential PCR product downregulated by the coculture of endothelial cells with tumor cells. We further demonstrated that the downregulation was due to the level of endothelial cell confluency (not shown). As shown in Fig 2, the sequence of the PCR product comprised a partial open reading frame containing the canonical sequence of C_2 domain: YQCxAS. It showed the C_2 canonical residues Asn and Gly at position +4 and +7 (bold) relative to the targeted cysteine. Furthermore, an additional cysteine residue was found at position +11 (circled), which corresponded to the situation in Titin or CTX sequences. The sequence terminated in a stretch of hydrophobic amino acids (underlined), which was consistent with a putative transmembrane region. Cloning of the full-length sequence was achieved by the 5'race technique, and the full-length sequence encoded for a type I protein containing a V and a C_2 domain. Since the sequence surrounding the cysteine conserved in the C_2 domain corresponded to the targeted sequence by the degenerated primers, this validated the criteria used to select the PCR products corresponding to putative novel Ig Sf molecules.

```
tat cgg tgc att gct tcc aat gac gca ggt gca gcc agg tgt gag ggg cag gac atg gaa :60
Y   Q   C   I   A   S   N   D   A   G   A   A   R  (C)  E   G   Q   D   M   E   :20
gtc tat gat ttg aac att gct ggg att att ggg gga gtc ctt gtt gtc ctt att gtt ctt :120
V   Y   D   L   N   I   A   G   I   I   G   G   V   L   V   V   L   I   V   L   :40
gct gtg att acg atg ggc atc tgc tgt gcg tac aga cga ggc tgc ttc atc agc agt aaa :180
A   V   I   T   M   G   I   C   C   A   Y   R   R   G   C   F   I   S   S   K   :60
caa gat gga gaa agc tat aag agc cca ggg aag cat gac ggt gtt aac tac atc cgg acg :240
Q   D   G   E   S   Y   K   S   P   G   K   H   D   G   V   N   Y   I   R   T   :80
agt gag gag ggt gac ttc aga cac aaa tcg tcc ttt gtt atc tga cac ctg tcg gct ggg :300
S   E   E   G   D   F   R   H   K   S   S   F   V   I   *   H   L   S   A   G   :100
aga gca cat gca agt acc tct gtt gga agc tgg aca ctg ata                         :342
R   A   H   A   S   T   S   V   G   S   W   T   L   I                           :114
```

Fig. 2: Sequence of one differential PCR product likely encoding a novel IgSf molecule. Sequences of the degenerated primers are indicated as bold nucleotidic sequences.

By comparing the protein sequence with nucleic acid databases, we identified the closest related sequences as JAM-1, and a mouse EST (Acc N°: AA445150), previously identified as a member of the CTX gene family (14). The transcript identified by differential display was named JAM-2; and the murine EST, JAM-3. The alignment of JAM-1, JAM-2 and JAM-3 is shown on Fig 3.

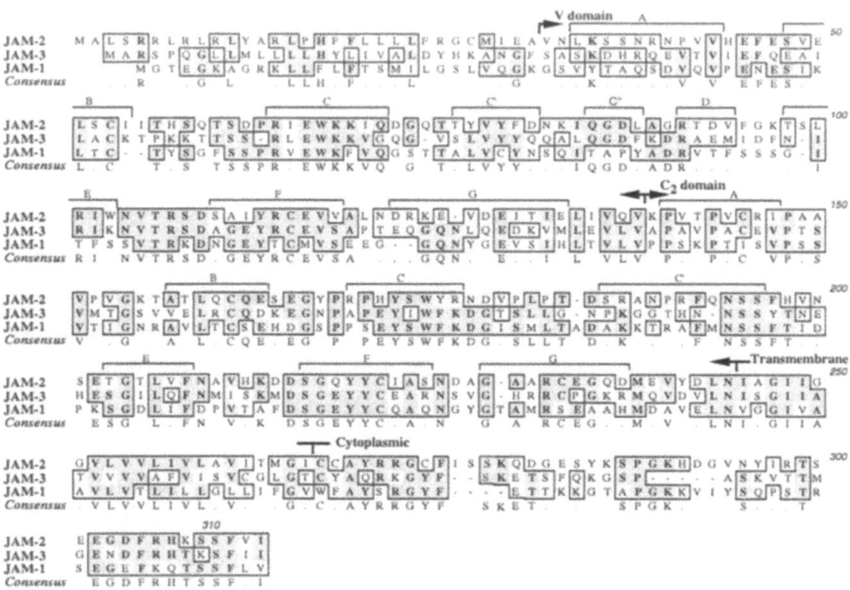

Fig. 3: Alignment of Junctional Adhesion Molecular family members.

The three proteins are type I proteins with two extracellular Ig Sf domains, a transmembrane and a cytoplasmic segment. The membrane distal domain is of V type, the membrane proximal domain is of the C_2 type. In JAM-2 and JAM-3 but not in JAM-1, the C_2 domain contains an extra pair of cysteines at the same position as in the CTX family members. This indicates that JAM-2 and JAM-3 are closer to the CTX family than JAM-1, but all three proteins seem to belong to a different subset of this family since their cytoplasmic parts are conserved in length and composition. Other features of this subset seem to be conserved in evolution: three molecules of zebra fish and *Xenopus* have been identified in the database. The C strand (PRV/IEWKF/K) of their V domain is remarkably similar to that of JAM-2, JAM-3 or JAM-1 (17). This suggests that the Junctional Adhesion Molecule subset of the CTX family may perform similar functions inside the cells once their external ligands are engaged. Their respective functions can be either the participation in intercellular contacts between endothelial cells and vascular cohesion, or the control of leukocyte emigration if the external ligand is present on transmigrating cells.

We directly tested this hypothesis by fusing JAM-2, JAM-3 or JAM-1 to EGFP and tracking the subcellular localization of the molecules (Fig 4). Interestingly, JAM-1-EGFP and JAM-2-EGFP were specifically localized to cell-cell contacts between transfected cells, but not between transfected and non-transfected cells. This indicated that common sequences between JAM-1 and JAM-2 may be sufficient to support the enrichment of the molecules to cell-cell contacts. Interestingly, JAM-3 was not enriched to cell-cell contacts, leading to the hypothesis that JAM-3 was not coupled to the same intracellular mediators as JAM-2 or JAM-1, and was unable to interact homophilically. Notably, the loss of specific localization of JAM-3 to cell-cell contacts is correlated with the absence of putative PKC phosphorylation sites in the JAM-3 cytoplasmic sequence. Nevertheless, we cannot exclude that more subtle differences in the V domain of JAM-3, compared to that of JAM-1 or -2, may explain the loss of localization of JAM-3 to contacts. Analysis of the amino-acid sequences responsible for such an enrichment of JAM-1 or JAM-2 to cell-cell contacts will improve our understanding of the participation of JAM-1 or JAM-2 to tight junctional complexes of polarized cells (5). Moreover, the comparison between the relative tissue distribution of JAM-1, JAM-2 and JAM-3 has shown that each member of the novel protein family present overlapping but distinct pattern of expression (not shown). JAM-2 was found specifically expressed by endothelial cells of high endothelial venules in secondary adult lymphoid organs, suggesting that it may play a role in the control of leukocyte recirculation.

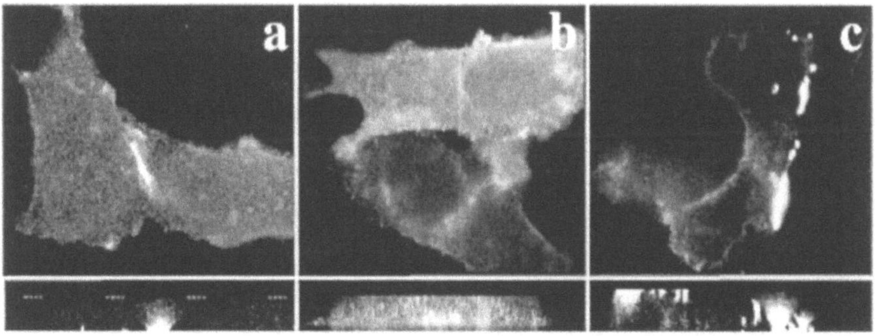

Fig. 4: Localization of (a) JAM-2-EGFP, (b) JAM-3-EGFP and JAM-1-EGFP (c) in CHO cells.

Perspectives and concluding remarks

JAM-2 was identified as a transcript which was downregulated in endothelial cells under the influence of tumors. We further showed that the transcriptional downregulation was due to the increased endothelial cell confluency under the coculture condition. JAM-2 is a member of a novel molecular family including JAM-3 and JAM-1, a recently described tight junctional molecule regulating leukocyte emigration and paracellular permeability (3, 13). Since we found JAM-2 specifically localized to tight junctions (not shown), one could ask whether the endothelial tight junctional complexes might be regulated during the angiogenic switch. Accordingly, endothelial migration and division observed during angiogenesis may imply modifications of endothelial cell polarity. More generally, little is known about the role and the redistribution of molecules participating in intercellular adhesive complexes during tissue remodeling. The characterization of novel surface molecules participating in tight junctions will increase our knowledge of the relationship between the intercellular crosstalk and the structure of intercellular adhesive complexes.

Acknowledgements

I would thank my colleagues, Claude Magnin and Dominique Ducrest for their invaluable help. The work was supported by Swiss National Science Foundation, Schweizerische Krebsliga and Human Frontier Science Program Organization to MAL. The Basel Institute was founded and is supported by F. Hoffmann-La roche, AG, Basel, Switzerland.

References

1. Flamme, I., T. Frolich, and W. Risau. 1997. Molecular mechanisms of vasculogenesis and embryonic angiogenesis. J Cell Physiol. 173:206-210.
2. Risau, W. 1997. Mechanisms of angiogenesis. Nature. 386:671-674.
3. Martin-Padura, I., S. Lostaglio, M. Schneemann, L. Williams, M. Romano, P. Fruscella, C. Panzeri, A. Stoppacciaro, L. Ruco, A. Villa, D. Simmons, and E. Dejana. 1998. Junctional adhesion molecule, a novel member of the immunoglobulin superfamily that distributes at intercellular junctions and modulates monocyte transmigration. J Cell Biol. 142:117-127.
4. Lampugnani, M.G., M. Resnati, M. Raiteri, R. Pigott, A. Pisacane, G. Houen, L.P. Ruco, and E. Dejana. 1992. A novel endothelial-specific membrane protein is a marker of cell-cell contacts. J Cell Biol. 118:1511-1522.
5. Bischoff, J. 1997. Cell adhesion and angiogenesis. J Clin Invest. 100:S37-S39.
6. Folkman, J. 1995. Angiogenesis in cancer, vascular, rheumatoid and other disease. Nat Med. 1:27-31.
7. Lewalle, J.M., K. Bajou, J. Desreux, M. Mareel, E. Dejana, A. Noel, and J.M. Foidart. 1997. Alteration of interendothelial adherens junctions following tumor cell- endothelial cell interaction in vitro. Exp Cell Res. 237:347-356.
8. Sabapathy, K.T., M.S. Pepper, F. Kiefer, U. Mohle-Steinlein, F. Tacchini-Cottier, I. Fetka, G. Breier, W. Risau, P. Carmeliet, R. Montesano, and E.F. Wagner. 1997. Polyoma middle T-induced vascular tumor formation: the role of the plasminogen activator/plasmin system. J Cell Biol. 137:953-963.
9. Brooks, P.C., R.A. Clark, and D.A. Cheresh. 1994. Requirement of vascular integrin alpha v beta 3 for angiogenesis. Science. 264:569-571.
10. Jain, R.K., G.C. Koenig, M. Dellian, D. Fukumura, L.L. Munn, and R.J. Melder. 1996. Leukocyte-endothelial adhesion and angiogenesis in tumors. Cancer and Metastasis Reviews. 15:195-204.
11. Griffioen, A.W. 1997. Phenotype of the tumor vasculature; cell adhesion as a target for tumor therapy. Canc J. 10:249-255.
12. Piali, L., A. Fichtel, H.-J. Terpe, B.A. Imhof, and R.H. Gisler. 1995. Endothelial vascular cell adhesion molecule 1 expression is suppressed by melanoma and carcinoma. J Exp Med. 181:811-816.
13. Malergue, F., F. Galland, F. Martin, P. Mansuelle, M. Aurrand-Lions, and P. Naquet. 1998. A novel immunoglobulin superfamily junctional molecule expressed by antigen presenting cells, endothelial cells and platelets. Mol Immunol. 35:1111-1119.
14. Chretien, I., A. Marcuz, M. Courtet, K. Katevuo, O. Vainio, J.K. Heath, S.J. White, and L. Du Pasquier. 1998. CTX, a Xenopus thymocyte receptor, defines a molecular family conserved throughout vertebrates. Eur. J. Immunol. 28:4094-4104.
15. Du Pasquier, L., M. Courtet, and I. Chretien. 1999. Duplication and MHC linkage of the CTX family of genes in Xenopus and in mammals. Eur J Immunol. 29:1729-1739.
16. Willams, A.F., and A.N. Barclay. 1988. The immunoglobulin superfamily - domains for cell surface recognition. Ann. Rev. Immunol. 6:381-405.
17. Du Pasquier, L. 1999. Relationships among the genes encoding MHC molecules and the specific antigen receptors. In Proceeding of the sixth workshop on MHC evolution. M. Kasahara, editor. Springer.

III

Thymus Development

Thymus Epithelial Cell Reaggregate Grafts

H.-R. Rodewald
Basel Institute for Immunology, Grenzacherstr. 487, CH-4005 Basel, Switzerland
Present address: Immunology, University of Ulm, Albert-Einstein-Allee 11, D-89081 Ulm, Germany

Introduction

In the thymus, a unique three-dimensional microenvironment fosters the development of T lymphocytes. T cell progenitors are attracted to the thymus from the circulation by enigmatic molecular and cellular mechanisms. Following colonization, thymocytes undergo a highly ordered developmental process involving several phenotypically and functionally distinct steps. First, signalling via growth factor receptors mediates proliferation and protection from growth factor withdrawal-mediated apoptosis, thus facilitating expansion of the small pool of pro-T cells ("growth factor expansion phase" [reviewed by DiSanto and Rodewald 1998]. Second, β, γ and δ T cell antigen receptor (TCR) genes are rearranged. Thymocytes expressing a productive TCR β chain assembled into the pre-TCR complex are selected for further development, a process accompanied by massive proliferation ("β selection") [von Boehmer and Fehling 1997]. Third, thymocytes replace the pre-TCR α chain by highly diverse TCR α chains, assemble complete αβ TCRs, and are selected based on the specificity of these receptors to yield self-tolerant ("negative selection"), MHC-restricted ("positive selection") T cells (collectively termed "αβ selection") [Kisielow and von Boehmer 1995]. Thymocytes undergo further differentiation as CD4 or CD8 single-positive cells before they can head as antigen-reactive T cells for the peripheral lymphoid organs.

During these developmental steps, thymocytes migrate through the various stromal cell sites (i.e. from the medulla to the cortex and back to the medulla). The specific functions of such "local units" of hematopoietic environments are only poorly understood, both for the thymus and for the bone marrow. Stromal cells can provide growth and differentiation signals via cell-cell contact (e.g. MHC – TCR) and via soluble factors. For instance, thymus-colonizing progenitors express cytokine receptors, including IL-7 receptors, and tyrosine kinase receptors, including c-kit. The ligands for these receptors are expressed by thymic epithelial cells. The unique function of the thymic epithelium is underscored by the fact that mutations affecting the thymus anlage, such as those caused by the winged-helix-nude (whn) [Nehls et al. 1994] mutation in mice or the DiGeorge syndrome in humans, can lead to ablation of the entire T lymphocyte lineage. Stromal cells, when defined as the cohort of cells and tissues supporting thymocyte development, can be of epithelial (thymus stroma "proper" forming cortical and medullary zones), mesenchymal (fibroblasts), and hematopoietic (dendritic cells, macrophages) origin. In addition, the thymus harbors endothelial cell-lined blood vessels, and neuronal cells [reviewed by von Gaudecker 1991].

Unlike B cell development which has been successfully studied in two-dimensional tissue culture systems [reviewed by Rolink et al. 1995], it has proven difficult to recapitulate the physiology of intrathymic T cell development in "simple" two-dimensional stromal cell cultures seeded with T cell progenitors. Although experiments have been reported [e.g. Dejbakhsh-Jones and Strober 1999] in which various steps of thymocyte development were promoted by two-dimensional stromal cell cultures, by and large, three-dimensional thymic stromal cell systems have proven superior, if not obligatory, to facilitate studies on T cell development *in vitro* [reviewed by Anderson et al. 1996]. Deficiencies of stromal cell monolayers in supporting T cell development *in vitro* may be explained by the fact that thymic stromal cells, maintained as monolayer cultures as opposed to reaggregated three-dimensional cultures (see below), alter their expression pattern of key genes such as MHC class II or the whn gene [Anderson et al. 1998]. Due to the limitations of two-dimensional cultures, fetal thymic organ cultures (FTOC), depleted of endogenous and repopulated by exogenous progenitors, have been widely used to study intrathymic development kinetically, and, to some extent, also quantitatively [reviewed by Jenkinson and Owen 1990].

Analyses of the role of subsets of the thymic stroma for the formation of a functional thymus environment became accessible only with the introduction of methods in which the fetal thymus was dissociated and reassembled *in vitro* [Anderson et al. 1993]. Such reaggregated fetal thymus organ cultures, designated RFTOC, have been used to study both stromal cell and progenitor requirements *in vitro* [e.g.Oosterwegel et al. 1997; Muller et al. 1997]. However, in these and other studies addressing cellular or molecular requirements of T cell development in RFTOC, it has been difficult to exclude the possibility that deficiencies in RFTOC formation *per se* led to altered or abrogated T cell development. This poses an inherent problem which has not been solved yet. Another deficiency of "*in vitro*-RFTOC" is due to the fact that the reaggregates can be cultured for a period of no longer than ~12 days. During this time, the RFTOC architecture does not resemble the normal thymic structure.

Here, experiments are summarized in which the capacity of RFTOC to function in lieu of the endogenous thymus following transplantation under the kidney capsule of recipient mice was studied. The reviewed results show that (1) RTFOC assembled *in vitro* from single cell suspensions of fetal thymic epithelium can attract host (bone marrow)-derived T cell progenitors from the circulation, (2) these pro-T cells develop along the well-defined stages of intrathymic T cell development, (3) the fidelity of negative and positive selection is apparently normal when examined in TCR-transgenic recipient mice, and (4) the random structure, present in RFTOC maintained *in vitro*, can "self-reorganize" into a proper thymic architecture with clear medulla-cortex boundaries after grafting.

Components of the Thymic Environment

The components of the thymic environment have been analyzed morphologically [reviewed by von Gaudecker 1991] and immunocytochemically [reviewed by Boyd et al. 1993; van Ewijk 1991] in great detail These analyses suggest high complexity, both in terms of cellular components and products (epithelium, fibroblasts, connective tissues, endothelial cells, neuronal cells) and architecture (capsule, septae, ordered blood vessel topology, cortical vs. medullary epithelial boundaries) of the thymic microenvironment. Panels of monoclonal antibodies, raised against subsets of stromal cells, detect in particular epithelial and connective tissue structures in the thymus [Boyd et al. 1993; van Ewijk 1991]. Since most of these reagents recognize their antigens in histological assays, but not by cell surface staining, these tools are, unfortunately, of limited use for the prospective purification of stromal cell subsets. Some reagents, however, such as the cortical marker CDR1 [Rouse et al. 1988], or the medullar marker A2B5 [Haynes et al. 1983] can be used to isolate subsets of the thymic epithelium [Anderson et al. 1993].

Studies on the Reconstruction of a Thymus Environment *in vivo*

Studies on the capacity of subsets of the thymic epithelium to support T cell development *in vitro* have been reported (see Introduction). However, experiments addressing the properties of RFTOC following grafting have, to my knowledge, not been published. Therefore, both the functional capacity as well as the structural features of RFTOC grafts will be discussed below. The basic experimental outline is summarized in Fig. 1.

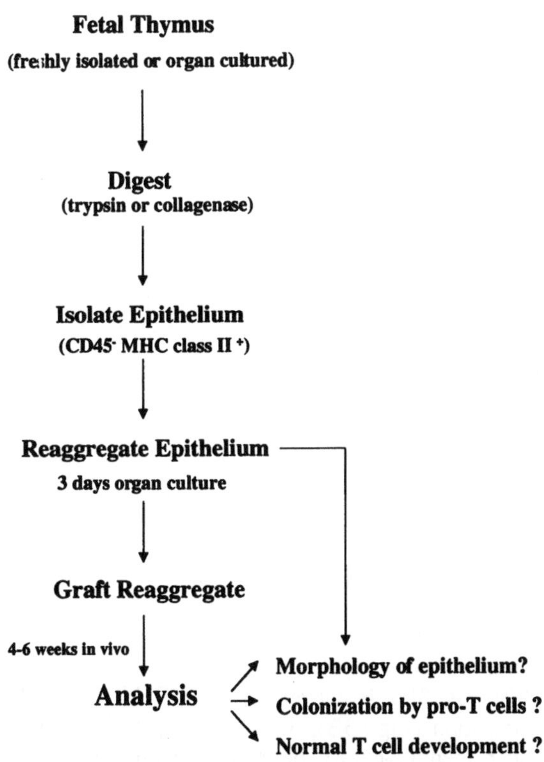

Fig. 1. Experimental outline for the reconstruction of a thymus environment *in vivo*

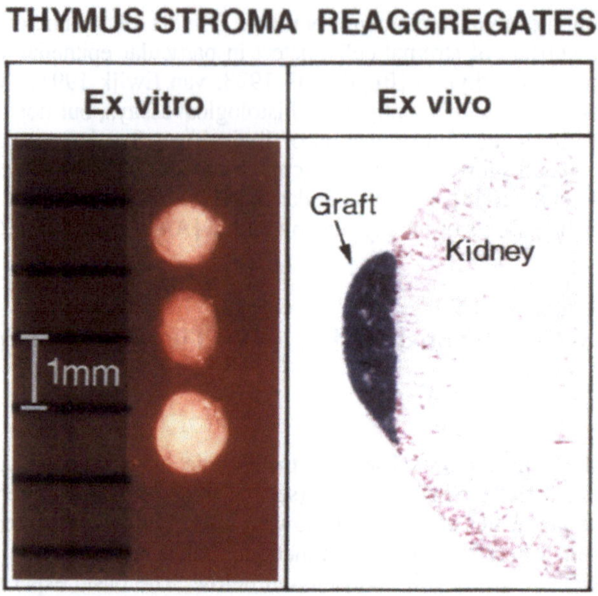

Fig. 2. Size and macroscopic appearance of alymphoid RFTOC before (left) and after (right) implantation under the kidney capsule. RFTOC were formed from ~200 nl of epithelial cell suspensions. Reggregates had a diameter of ~1mm after 2-3 days in culture (left). 4 weeks after transplantation, histological analysis revealed that RFTOC resembled a lymphoid tissue according to May-Grünwald-Giemsa staining (right).

Self-Organization of Epithelial Structures in RFTOC Grafts

Thymic epithelium isolated on days 15.5 or 16.5 of fetal development was purified positively by cell sorting (CH45$^-$ MHC class II$^+$ or other stromal cell phenotypes, e.g. CDR1$^+$ A235$^-$), or negatively by elimination of hematopoietic cells via CD45-depletion using magnetic beads. Thymic epithelium was reassembled *in vitro* to form reaggregated fetal thymus organ cultures as described [Anderson et al. 1993]. Size and macroscopic appearance of alymphoid RFTOC before and after implantation under the kidney capsule are show in Fig. 2. RFTOC formed from ~200 nl volume of epithelial cell suspension yielded reggregates with a diameter of ~1mm after 2-3 days in culture (Fig. 2, left panel). From 4 weeks onward following transplantation, histological analysis revealed that RFTOC resembled a lymphoid tissue according to intense May-Grünwald-Giemsa staining (Fig. 2, right panel). Grafts were analyzed immunocytochemically with a variety of markers specific for cortical (MTS44 [Boyd et al. 1993]) and medullary (MTS10 [Boyd et al. 1993], UEA-1 [Farr and Anderson 1985] epithelium, septae (MTS16 [Boyd et al. 1993]), as well as MHC class II of donor and host origin. In normal thymic tissue sections, medullary and cortical epithelium are MHC class II^{high} and MHC class II^{low} respectively [reviewed by van Ewijk 1991] (Fig. 3A). MHC class II expression was maintained when RFTOC were kept *in vitro* [Anderson et al. 1993; Anderson et al. 1998] (Fig. 3B). However, *in vitro*, RFTOC lacked any noticeable medulla-cortex organization (Fig. 3B). Interestingly, MHC class II and UEA-1 staining revealed that medulla-cortex boundaries were re-established following implantation (Fig. 3 C,D).

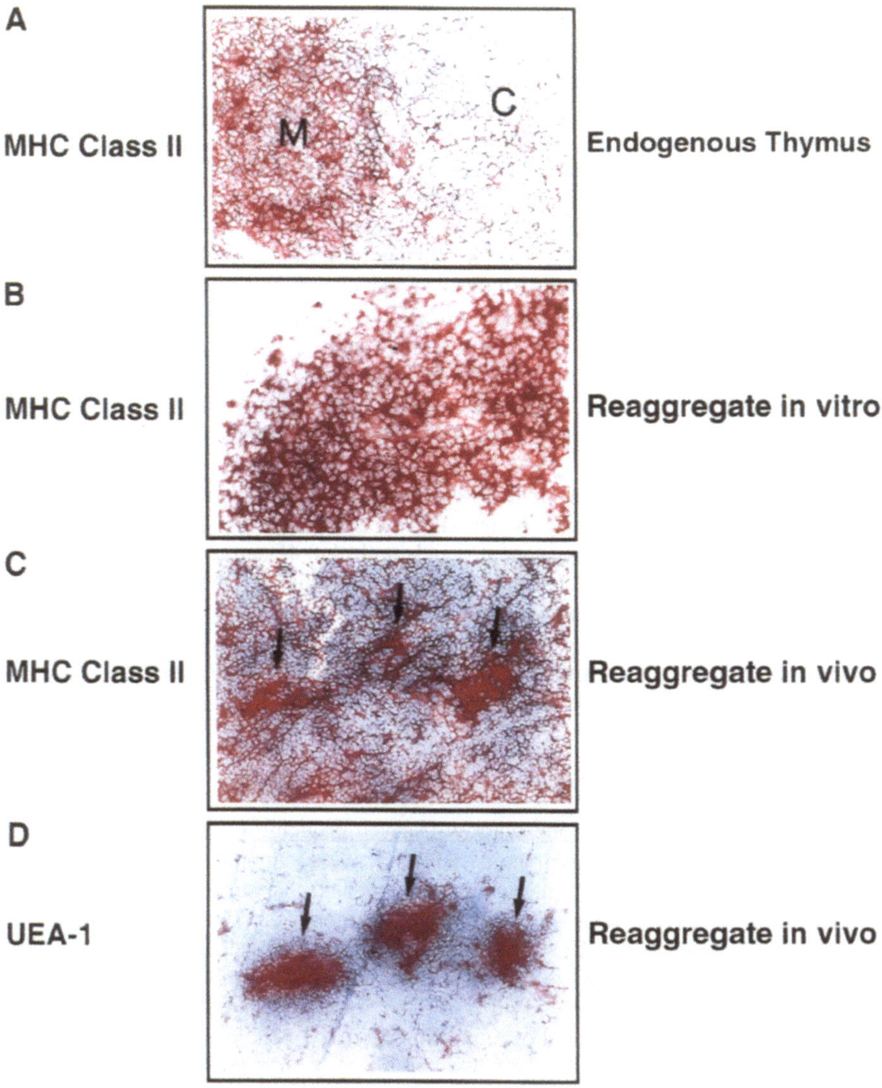

Fig. 3. Epithelial reorganization following transplantation of fetal thymic stromal cell reaggregates. Tissue sections from normal thymus (A), reaggregated thymic epithelial grafts maintained *in vitro* (B) or following transplantation under the kidney capsule (C, D) were stained with MHC class II-specific mAb (A, B, C) or with the medullary marker UEA-1 [Farr, Anderson 1985] (D). In the normal thymus, medullary and cortical epithelium express MHC class II strongly and weakly, respectively (A). In reaggregates kept *in vitro*, MHC class II expression is maintained at high levels [Anderson et al. 1993] (B). Following grafting of RFTOC, the thymic epithelium is reorganized showing a marked medulla-cortex organisation as visualized by staining for MHC class II (arrows in C) or the medulla specific marker UEA-1 (arrows in D). Both markers recognize three identical medullary "islets" within the cortex on serial sections.

To determine whether this *in vivo* reorganization of medulla-cortex organization occurred either via segregation and clustering (sorting-out) of pre-existing medullary epithelial cells or, alternatively, from clonal progenitors, mixing experiments were performed. To this end, RFTOC were assembled *in vitro* from two mouse strains differing in their MHC. A 1:1 mixture of C57Bl/6 (I-A^b) and Balb/c (I-E^d) fetal thymic epithelium was used to 'build' RFTOC *in vitro.* To avoid colonization of MHC class II expressing cells (dendritic cells, macrophages) from the host, such mixed reaggregates were implanted into MHC class II deficient C57Bl/6 mice [Kontgen et al. 1993]. In such MHC class II^+ grafts, mature $CD4^+CD8^-$ thymocytes developed indicating that the graft could facilitate positive selection (see below). As already stated, for single-strain-derived RFTOC, medullary areas, clearly distinct from cortical areas, were observed following transplantation. Surprisingly, double-staining with anti-I-A^b vs. anti-I-E^d demonstrated that individual medullary islets were derived from either of the two donors, but not from both. In contrast, the cortex was composed of epithelial cells from both donor types.

Embryological analyses had previously suggested that medulla-cortex compartmentalization occurs via invagination of an endodermal into an ectodermal epithelial sheet at the third pharyngeal pouch and cleft [Benoist and Mathis 1999; Picker and Siegelman 1999]. In contrast to this view, the experiments outlined here suggest that thymus medullary epithelium is not composed of an epithelial layer. Rather, based on the single-donor origin of medullary areas, the medullary epithelium appears to be composed of individual epithelial islets, each of which is derived from a single epithelial stem cell [HR Rodewald, C. Haller, SS Tan, unpublished].

Functional Properties of Thymic Reaggregate Grafts

Graft Colonization by Host Bone Marrow-Derived Progenitors

When B6Ly5.1-derived, alymphoid RFTOC were implanted into B6Ly5.2 congenic recipients, flow cytometric analysis revealed that stromal cell grafts were colonized by host type, bone marrow-derived progenitors. Analysis for the host bone marrow marker vs. CD4, CD8 or $\alpha\beta$ TCR demonstrated that the graft-colonizing cells gave rise to the normal pattern of intrathymic T cell development. Thus, the microenvironment provided by the reorganized graft was capable of attracting T cell progenitors, and of promoting normal T cell development *in vivo.*

Negative Selection of Self-Reactive Thymocytes in the Grafts

Since RFTOC architecture closely resembled the normal thymic structure, the next set of experiments addressed the question whether RFTOC would select thymocytes according to the rules of positive and negative selection [reviewed by Kisielow and von Boehmer 1995]. Negative selection was examined in the TCR transgenic mouse expressing an H-Y (male) specific, H-2 D^b restricted TCR [von Boehmer 1990]. Since thymi were harvested from large numbers (50-100) of fetal mice which were not separated according to their sex, and since epithelial cells were pooled from all thymi, RFTOC were assembled from approximately 50% male and 50% female epithelial cells. When such male/female mixed, H-2^b RFTOC were implanted into female H-Y transgenic mice, the thymocyte phenotype was indicative of ongoing massive negative selection, i.e. the fraction of $CD4^+CD8^+$ thymocytes was strongly reduced. The phenotype of thymocytes retrieved from RFTOC grafts under conditions of negative selection was identical to the phenotype known from male H-2^b H-Y TCR transgenics [von Boehmer 1990]. Thus, based on this particular MHC class I restricted TCR, the fidelity of negative selection of autoreactive TCRs appears to be normal in RFTOC *in vivo*.

Positive Selection of Self-MHC-Restricted Thymocytes in the Grafts

In analogy to the negative selection experiments discussed above, RFTOC were also examined for their capacity to promote positive selection of thymocytes based on TCR/H-2 restriction. To this end, thymic epithelium was isolated from H-2^d mice, reaggregated *in vitro*, and grafted into a TCR transgenic mouse expressing an I-E^d restricted, influenza virus hemagglutinin-specific TCR on a RAG-2-deficient background (HA-TCR mice) [Kirberg et el. 1994]. These mice select $CD4^+CD8^-$ clonotype$^+$ thymocytes in the context of H-2^d but not H-2^b. To test positive selection in RFTOC *in vivo*, thymic epithelium was isolated from H-2^d mice, and implanted into H-2^b HA-TCR mice. The endogenous thymus of these recipient mice was, as expected, unable to generate $CD4^+CD8^-$clonotype$^+$ thymocytes. In contrast, H-2^d expressing grafts selected $CD4^+CD8^-$clonotype$^+$ thymocytes, and a fraction of these cells was peptide antigen responsive when analysed *in vitro*.

Collectively, RFTOC grafts behaved indistinguishably from either endogenous thymi or intact thymic grafts when analysed under conditions of negative or positive selection. Since T cell development and selection are closely controlled by the thymic environment, the self-reorganized structure of the thymic epithelial reticulum is also functionally fully reorganized. It is conceivable that even minor structural alterations affecting niche size, stromal cell density or stromal cell position within the thymus might affect the selection process. Thus, it appears that RFTOC grafts fulfil the structural requirements for a functional thymus *in vivo*.

Conclusions

Transplantation of thymic epithelial reaggregate grafts revealed the following information: (1) RFTOC are colonized *in vivo* by bone marrow-derived progenitors. (2) Pro-T cells developing in the grafts follow the well-defined pathways of intrathymic T cell development. (3) Thymocytes developing in these grafts obey the rules of positive and negative selection when grafts are implanted into transgenic mice expressing TCRs which are either self-reactive (H-Y specific, H-2 D^b restricted TCR) or non-self-reactive (influenza virus hemagglutinin-specific, I-E^d restricted). (4) Remarkably, the thymic epithelium undergoes tissue-reorganization following grafting such that medullary and cortical epithelium, localized randomly due to tissue digestion and *in vitro* reassembly, are segregating from each other to form distinct, sharply demarcated medullary and cortical zones. This system may yield additional insight into the molecular and cellular requirements for the formation and reformation of functional thymic tissue *in vivo*.

Acknowledgements

I thank Corinne Haller and Birgit Kugelberg for expert technical assistance, and F. Melchers and S. Gilfillan for discussions. The Basel Institute for Immunology was founded and is supported by F. Hoffmann-La Roche, Basel, Switzerland.

References

Anderson G, Jenkinson EJ, Moore NC, Owen JJ (1993) MHC class II-positive epithelium and mesenchyme cells are both required for T-cell development in the thymus. Nature 362:70-73

Anderson G, Moore NC, Owen JJT, Jenkinson EJ (1996) Cellular interactions in thymocyte development. Annual Review of Immunology 14:73-99

Anderson KL, Moore NC, McLoughlin DE, Jenkinson EJ, Owen JJ (1998) Studies on thymic epithelial cells in vitro. Dev Comp Immunol 22:367-77

Benoist C, Mathis D (1999) T-Lymphocyte Differentiation and Biology. In: W. Paul (ed.) Fundamental Immunology. Lippincott-Raven, Philadelphia, New York

Boyd RL, Tucek CL, Godfrey DI, Izon DJ, Wilson TJ, Davidson NJ, Bean AG, Ladyman HM, Ritter MA, Hugo P (1993) The thymic microenvironment. Immunology Today 14:445-459

Dejbakhsh-Jones S, Strober S (1999) Identification of an early T cell progenitor for a pathway of T cell maturation in the bone marrow. Proc Natl Acad Sci U S A 96:14493-8

DiSanto JP, Rodewald H-R (1998) In vivo roles of receptor tyrosine kinases and cytokine receptors in thymocyte development. Current Opinion in Immunology 10:196-207

Farr AG, Anderson SK (1985) Epithelial heterogeneity in the murine thymus: fucose-specific lectins bind medullary epithelial cells. Journal of Immunology 134:2971-2977

Haynes BF, Shimizu K, Eisenbarth GS (1983) Identification of human and rodent thymic epithelium using tetanus toxin and monoclonal antibody A2B5. J Clin Invest 71:9-14

Jenkinson EJ, Owen JJ (1990) T-cell differentiation in thymus organ cultures. Semin Immunol 2:51-8

Kirberg J, Baron A, Jakob S, Rolink A, Karjalainen K, von Boehmer H (1994) Thymic selection of $CD8^+$ single positive cells with a class II major histocompatibility complex-restricted receptor. J Exp Med 180:25-34

Kisielow P, von Boehmer H (1995) Development and selection of T cells: facts and puzzles. Advances in Immunology 58:87-209

Kontgen F, Suss G, Stewart C, Steinmetz M, Bluethmann H (1993) Targeted disruption of the MHC class II Aa gene in C57BL/6 mice. International Immunology 5:957-964

Muller KM, Luedecker CJ, Udey MC, Farr AG (1997) Involvement of E-cadherin in thymus organogenesis and thymocyte maturation. Immunity 6:257-64

Nehls M, Pfeifer D, Schorpp M, Hedrich H, Boehm T (1994) New member of the winged-helix protein family disrupted in mouse and rat nude mutations. Nature 372:103-107

Oosterwegel MA, Haks MC, Jeffry U, Murray R, Kruisbeek AM (1997) Induction of TCR Gene Rearrangements in Uncommitted Stem Cells by a Subset of IL-7 Producing, Class II–Expressing Thymic Stromal Cells. Immunity 6:351-360

Picker LJ, Siegelman MH (1999) Lymphoid Tissues and Organs. In: W. Paul (ed.) Fundamental Immunology. Lippincott-Raven, Philadelphia, New York

Rolink A, Ghia P, Grawunder U, Haasner D, Karasuyama H, Kalberer C, Winkler T, Melchers F (1995) In-vitro analyses of mechanisms of B-cell development. Semin Immunol 7:155-67

Rouse RV, Bolin LM, Bender JR, Kyewski BA (1988) Monoclonal antibodies reactive with subsets of mouse and human thymic epithelial cells. J Histochem Cytochem 36:1511-1517

van Ewijk W (1991) T-cell differentiation is influenced by thymic microenvironments. Annual Review Immunology 9:591-615

von Boehmer H (1990) Developmental biology of T cells in T cell receptor transgenic mice. Ann Rev Immunol 8:531-556

von Boehmer H, Fehling HJ (1997) Structure and function of the pre-T cell receptor. Annu Rev Immunol 15:433-52

von Gaudecker B (1991) Functional histology of the human thymus. Anat Embryol(Berl) 183:1-15

A Novel Anti-Ep-CAM Antibody to Analyze the Organization of Thymic Medulla in Autoimmunity

M. Naspetti[1], F. Martin[1], A. Biancotto[1], F. Malergue[1], P. Mansuelle[3], F. Galland[1,2] and P. Naquet[1,2]

[1] Centre d'Immunologie de Marseille-Luminy INSERM U136-CNRS UMR145, Marseille, France

[2] Université de la Méditerranée, Faculté des Sciences de Luminy, Marseille, France

[3] UMR 6560 CNRS, Laboratoire de Biochimie et Ingéniérie des protéines, IFR J. Roche, Faculté de Médecine Nord, Marseille, France

Introduction

Organization of microenvironments through which lymphocytes migrate is crucial for the generation of optimal immune responses and development [1]. In peripheral lymphoid organs, chemokines secreted by stromal cells are required for the constitution of T and B cell areas [2]. In the thymus, immature thymocytes attach to fibronectin via of $\alpha5\beta1$ integrins [3]. Then, upon ligation of the pre-TCR, they upregulate CCR9 which triggers their attraction towards TECK-producing epithelial cells in the cortex and the medulla [4, 5]. Positively selected thymocytes lose CCR9 and acquire CCR4 and 7 which condition their migration towards chemokine gradient secreted by medullary stromal cells [6]. In relB-deficient mice, there is a complete disorganization of the cortico-medullary structure associated with a poor negative selection process [7-9]. These mice lack activated epithelial and myeloid-derived dendritic cells and, consequently, the expression of costimulatory or MHC molecules and the secretion of the medullary chemokine TCA4 are very low [10]. This animal model provides thus a useful link between the organization of a functional niche for negative selection and the activation of stromal cells. Using a panel of mAb directed at various types of stromal cells, we have found that the reactivity of some mAb was severely reduced in relB-deficient thymuses. One of them was found to recognize the 106/JAM molecule, a novel member of the IgSF highly expressed by endothelial cells and most MHC class II-expressing stromal cells in thymus and peripheral lymphoid organs [11]. This molecule plays a stabilizing role in the formation of tight junctions [12] and is acquired upon tissue colonization by macrophages and dendritic cells. In this report, we describe the molecular cloning of the Ep-CAM molecules identified by the mAb 29 [8] used to define relB-dependent thymic medullary stromal cells in normal and autoimmune-prone animals.

Material and Methods

Cell Lines and Antibodies

All the thymic stromal cell lines and antibodies 106, 29, 95, G8.8, 82B used in this study have been previously described [8, 11, 13, 14]. Others were obtained from Pharmingen (CD11c, CD11b, CD40, B7-1, B7-2).

Cloning of the Ep-CAM2 and Ep-CAM cDNA

The immunopurification of the molecule 29 from a MTE1D cell lysate and N-terminal peptide sequencing were performed as described [11]. Cloning of the 355 bp cDNA fragment from MTE1D mRNAs was done with the reverse (5257: 5'-GTTCAATGATGATCCAGTAG-3') and forward primers (5258 : 5'-CTGTGACAACTACAAGCTGG-3'). 3' RACE (6094 (5'-AGCGAGACCTGAG GGAGCTT-3')-UAP and 5' RACE (first strand primer 5257, (dC)n tailing and then use of the following reverse primers: 6179 (5'-GAAAGCTCCCTCAGGTCT CG-3') and 6256 (5'-CCAATTGAAGTACATTCGCAC-3') with the supplied AP primer) experiments were performed according to the manufacturer's advice.
The Ep-CAM clone was obtained by screening a grided E15 fetal thymic library with a probe corresponding to the 355bp 29 PCR fragment. This cDNA containing a full length Ep-CAM sequence was subcloned in the pRcCMV vector for transfection studies [13]. Northern and PCR analyses were performed according to previously described methodologies using the oligonucleotides 6094-6869 and 6870-6871 for Ep-CAM2 (200bp) and Ep-CAM (100bp) respectively (6094: AGAGCGAGACCTGAGGGAGCTTTCC, 6869: TGAAACACATAGTATTCCACA), (6870: CGGAGATCACGTGCTCCGAG, 6871: GAATATAGCTCTACACTTGC).

Phenotypic Analysis of Thymic Stromal Cells

Thymic stromal cells (10 thymus/exp) were dissociated according to published protocols [11]. After recovery from the Percoll gradient, the dissociated cells were not further deprived of lymphoid cells [15] and analysed by flow cytometry (FACScalibur, Becton Dickinson). Intrathymic injection of PKH26-labeled epithelial cell lines was performed through the thorax of anesthetized mice and immunohistochemistry with mAb 95 done on thymuses recovered 3 days later.

Results

The Antibody 29 Recognizes a Novel Ep-CAM2 Molecule

The 29 mAb was produced by immunizing rats with the MTE1D cell line and shown to immunoprecipitate a 50kD protein under non-reducing conditions and a dimer of 40 and 14-17kD polypeptides after reduction [8]. The immunopurified polypeptides were microsequenced and the sequences were: RACICENYKLTTxSLNINNQxExTSIGAQNSVICSKLATKxLVM, GLFKA, RARPEGAFQNTDGLYPD, and ERDRTYWIxIQLKHKTRExPYD. Homology searches in genome databases aligned them with mouse and human Ep-CAM-related sequences [16, 17]. However, the presence of several residue changes suggested that we might have identified a novel member of the Ep-CAM family in mouse. A combined strategy involving internal PCR and 3' and 5' RACE-PCR allowed us to get most of the coding sequence with the exception of that corresponding to the first three amino-acid residues. Unfortunately, the complete cDNA could not be obtained despite the screening of a genomic and several cDNA libraries. The available nucleotidic sequence has been submitted to GenBank and the corresponding peptidic sequence (calculated MW35kDa) is shown aligned with the Ep-CAM sequence (Genbank M76124) in Fig. 1.

```
EpCAM2   ...PQVLACGLLLAAATAAVAVAQRACICENYKLTTNCSLNINNQCECTSIGAQNSVICS   57
EpCAM1   MAGPQALAFGLLLAVVTATLAAAQRDCVCDNYKLATSCSLNEYGECQCTSYGTQNTVICS   60

EpCAM2   KLATKCLVMKAEMTGTKSGRRARPEGAFQNNDGLYDPDCDEKGLFKAKHGNGTTTCWC.N  116
EpCAM1   KLASKCLAMKAEMTHSKSGRRIKPEG.IQNNDGLYDPDCDEQGLFKAKQCNGTATCWCVN  119

EpCAM2   TAGVRRTDKDTEISCTERVRTYWIIIELKHKTRETPYDTQSLQNALKETLKNRYQLDPKY  176
EpCAM1   TAGVRRTDKDTEITCSERVRTYWIIIELKHKERESPYDHQSLQTALQEAFTSRYKLNQKF  179

EpCAM2   ITNILYENDLITIDLMQNSSQKAQNDVDIADVAYYFEKDVKDESLFHSCK.MDLRVNGEQ  235
EpCAM1   IKNIMYENNVITIDLMQNSSQKTQDDVDIADVAYYFEKDVKGESLFHSSKSMDLRVNGEP  239

EpCAM2   LDLDPGRTAIYYVDEKPPEFSMQGLQAGIIAVIVVVTLAVIAGIVVLVISRKNRMAKYEK  295
EpCAM1   LDLDPGQTLIYYVDEKAPEFSMQGLTAGIIAVIVVVSLAVIAGIVVLVISTRKKSAKYEK  299

EpCAM2   AEIKEMGEMHRELNA   310
EpCAM1   AEIKEMGEIHRELNA   314
```

Fig. 1. Alignment of Ep-CAM and Ep-CAM2 Peptidic sequences

As deduced from this alignment, the novel Ep-CAM2 sequence is more homologous to the human sequence GA733.2 (81% at the protein level) than to the murine Ep-CAM sequence (77.2%) suggesting that this novel sequence might be another mouse equivalent of the human GA733.2 sequence. All the critical domains of the Ep-CAM molecule are conserved in the Ep-CAM2 sequence including the EGF-like domain (aa 26-58), the thyroglobulin-like domain (aa 96-136) and the transmembrane domain (aa 267-288). Most of the cysteine residues are in conserved positions, mainly in the sequence stretch showing some homologies with nidogen (aa 26-142). In contrast, in the putative intracellular domain, the α-actinin binding domain present in the Ep-CAM molecule is modified in the Ep-CAM2 sequence leading to the loss of the dilysine motif (aa 299-293) involved in the interaction with the cytoskeletal protein [18].

Expression of Ep-CAM and Ep-CAM2 on Cell Lines and in Mouse Tissues

Since the mAb 29 recognized a novel Ep-CAM molecule, we wondered whether Ep-CAM might also be expressed by the MTE1D cell line and whether the 29 mAb would recognize Ep-CAM. For that reason, an Ep-CAM cDNA clone was fished out of a fetal day 15 thymic cDNA library and subcloned in the pRcCMV vector for transfection. Ep-CAM transfectants of the MTE4-14 cell line expressed both the 29 and Ep-CAM-specific G8.8 epitopes (RFI 40 in the transfected versus 3 in the untransfected cell line). In contrast, the G8.8 mAb did not stain the 29^{+} MTE1D cell line (data not shown). Furthermore, both mAb competed for physically linked epitopes on Ep-CAM transfectants (data not shown). To test whether the lack of reactivity of the G8.8 mAb was not due to post-transcriptional modifications of the Ep-CAM2 molecule, we performed a PCR analysis using oligonucleotides that specifically amplify Ep-CAM or Ep-CAM2-related sequences (Fig. 2). The results confirmed by the northern blot indicate that the MTE1D cell line does not express the Ep-CAM sequence (Fig. 2A,C). A larger analysis of the expression of both genes during thymic development and in mouse tissues reveals that the 29 transcript can only be detected in the MTE1D cell line whereas the Ep-CAM transcript is found in thymus, kidney, stomach and liver. Further PCR analysis using isoform-specific primers shows that the Ep-CAM2 transcript is undetectable during fetal thymic development whereas Ep-CAM is expressed in whole E15 or adult thymus and isolated thymocytes or in thymuses from RAG-deficient mice. Since the expression of the Ep-CAM2 gene is only detectable in the MTE1D cell line, we assume that the reactivity of the 29 mAb on mouse cryosections is directed towards the Ep-CAM molecule.

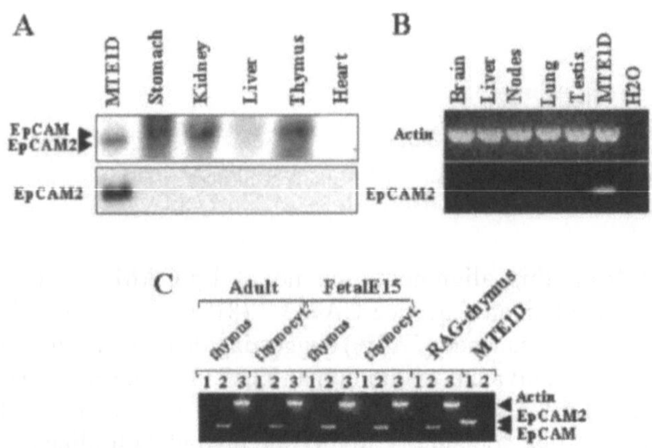

Fig. 2. Tissue distribution of EpCAM and EpCAM2 transcripts analyzed by: A) northern blot with an EpCAM (top panel) or an EpCAM2 (bottom panel) probe; B) and C) RT-PCR using isoform-specific primers on tissues on cell lines using actin as a control. C) 1, 2 and 3 correspond to EpCAM2, EpCAM and actin primers respectively.

Phenotypic Analysis of Ep-CAM^{+} Cells in Thymus

The mAb 29 recognizes a subset of UEA-1^{+} medullary stromal cells [8]. During thymic development, its reactivity is maximum on thymocytes and stromal cells in fetal day 15 thymuses and is progressively restricted to medullary stromal cells after fetal day 17. A low expression of Ep-CAM on most thymic epithelial [16] and some dendritic cells [14] has been reported. To further establish the cellular reactivity of the mAb 29, a flow cytometric analysis of dissociated thymic stromal cells was performed. Using indirect staining with biotinylated mAb 29 or G8.8, around 60% cells express moderate levels of the 29 epitope confirming that Ep-CAM is expressed at a low level on a large fraction of stromal cells. Among them, 21 % express Ep-CAM at a high level and using double staining with the CD11c marker, these Ep-CAM^{high} cells are split into CD11c$^+$ and CD11c$^-$ populations representative of dendritic and epithelial cells respectively (Table 1). Most 29^{high} epithelial cells express the CD54, CD40, CD80, CD86 and MHC class II antigens but not the CD11b monocyte marker. These cells overlap largely with the cell population defined by the anti-JAM 106 mAb previously shown to stain most MHC class II$^+$ cells in the thymus. These results confirm the notion that Ep-CAM is expressed at a high level by a subset of activated epithelial or dendritic cells in the thymic medulla. In addition to dendritic cells which are stained in T cell areas of lymph nodes, we found a staining of follicular dendritic cells in B cell follicles (data not shown).

Table 1: Relative proportions of stromal cell subsets in different mouse strains

Mice	mAbs	%[c]	Fraction of cells expressing the following markers[a]						
			106	CD11b	CD40	CD80	CD86	CD54	MHCII
BALB/c	29 high	13	62	0	45	25	9	50	5
	CD11c+	25	40	5	60	3	2	54	41
	DP[b]	9	78	14	50	60	37	47	56
NZB	29 high	4	47	0	15	3	0	64	9
	CD11c+	13	22	5	6	4	4	60	56
	DP	4	62	34	18	60	73	77	79
NOD	29 high	17	45	0	18	11	3	49	0
	CD11c+	44	23	6	34	2	2	51	0
	DP	11	70	8	44	66	43	70	49

[a] Staining obtained using a combination of 29-FITC, CD11c-PE and biotinylated mAb revealed by a SAV-APC probe (mean of 10 exp for BALB/c, 2-3 exp for NZB and NOD mice)
[b] 29^{high} and CD11c$^+$ cells
[c] Fraction of positive cells gated on high FSC and SSC values, thus excluding thymocytes

Ep-CAM$^+$ Thymic Stromal Cells in Autoimmune Mice

The development of a stromal cell population expressing high levels of Ep-CAM is dependent upon the expression of the relB transcription factor. Since the thymusus of NZB and NOD autoimmune-prone mice show signs of disorganization of the thymic medulla which precede the appearance of disease [19-21], we wondered whether they might show altered numbers of Ep-CAM$^+$ stromal cells. As seen in Table 1, the relative proportion of 29$^+$ and CD11c$^+$ stromal cells is reduced in NZB mice but not in 6-week-old NOD mice. Furthermore in NZB and NOD mice, one observes a reduction in the proportion of stromal cells expressing markers of activation such as CD80, CD86, MHC class II and CD40, confirming that the level of differentiation and/or activation of the thymic medulla is affected.

The MTE1D Cells Contribute to the Organization of the Thymic Medulla
The persistence of an organized thymic medulla (detected by the clustering of 95^+ medullary epithelial cells) in $relB^{-/-}$ bone marrow chimeras [8] suggested that in the absence of relB-dependent dendritic cells, the radioresistant 29^+ epithelial cells might be directly involved in the organization of 95^+ cells. As shown in Fig. 3, the injection of the PKH26-labelled MTE1D cell line into partially disorganized NZB thymuses is associated with a compact clustering of 95^+ cells around the injection point where the labelled injected cells remain tightly grouped. In contrast, the Ep-CAM^- epithelial cell line (MTE3-19) or its Ep-CAM transfectant (not shown) scatters after injection and does not attract 95^+ cells, suggesting that the Ep-CAM molecule is not sufficient to trigger this process. This indicates that the MTE1D cell line might recruit via cytokines other epithelial cells in thymic medulla.

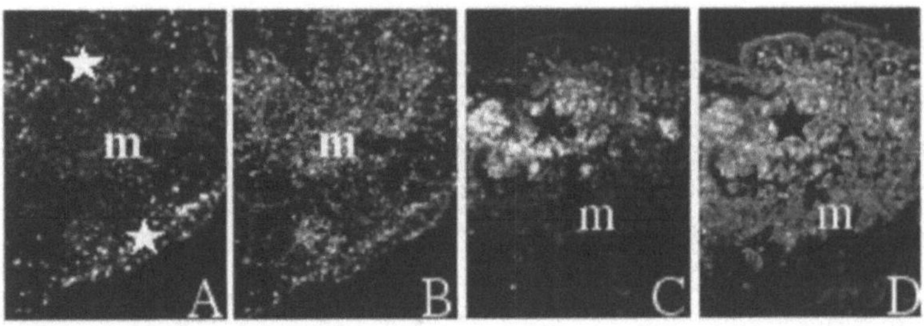

Fig. 3. Intrathymic injection (star) of PKH26-labeled MTE3-19 (A) or MTE1D (C) cells analysed on thymic cryosections counterstained with the mAb 95-FITC (B,D) showing medullary epithelium (m). Selection of a representative image.

Discussion

We describe the reactivity of a novel antibody to the Ep-CAM molecule and the identification of a second member of the mouse Ep-CAM family. Ep-CAM^+ epithelial and dendritic cells represent activated stromal cells in thymic medulla and their proportion is reduced in the NZB and NOD thymuses.
The sequence of the Ep-CAM2 molecule is highly homologous to the GA733.2 sequence coding for the human Ep-CAM molecule [22-24] and to the previously reported mouse Ep-CAM sequence [25]. The main differences are found in the α-actinin binding domain of the intracytoplasmic tail [26]. A Southern analysis using Ep-CAM or Ep-CAM2 probes allowed the detection of at least two restriction fragments but did not permit the identification of isoform-specific bands, making the possibility of a strain-related polymorphism unlikely (data not shown). These results suggest the existence of two independent genes in mouse. The northern and RT-PCR analysis confirmed that Ep-CAM was expressed in several tissues including thymus. Surprisingly, we failed to detect Ep-CAM2 transcripts in mouse, apart from the MTE1D cell line. Analysis using the two anti-

Ep-CAM mAb confirmed that G8.8$^+$ cells express the epitope 29 in vivo, whereas the MTE1D cell line only expresses Ep-CAM2. Thus the regulation of expression of the Ep-CAM2 gene is dissociated from that of the Ep-CAM gene and the reactivity of the mAb 29 in vivo is essentially due to the expression of the Ep-CAM molecule. The current and previous analyses of mice using the mAb 29 [8] agree with other reports using the G8.8 mAb [14, 16]. In addition, we found the expression of the Ep-CAM molecule by follicular dendritic cells. Interestingly, it has been previously reported that Ep-CAM is expressed by plasma cells, suggesting that it might participate in B lymphocyte interactions with FDC [25]. In the thymus, the mAb 29 identifies activated dendritic and epithelial cells which express high levels of Ep-CAM. These cells, absent in relB-deficient mice, express elevated levels of costimulatory, MHC class II molecules and chemokines. Their proportion and/or activation state are significantly reduced in the thymuses of NOD or NZB mice which display important alterations in the organization of the stroma as in relB$^{-/-}$ mice [19, 21]. These results suggest that a defect in stromal cell activation might be associated with the development of autoimmunity maybe by affecting negative selection. Ep-CAM acts as a homophilic cell adhesion molecule in man [27] and its ability to trigger stable adhesion requires a cytoplasmic interaction with α-actinin [26] in a region modified Ep-CAM2 molecule. The overexpression of Ep-CAM modulates cell proliferation, leads to the abrogation of adherens type junctions [18] and is frequently observed in carcinomas [28, 29]. Thus, this relB-linked enhanced expression might modulate the ability of 29$^+$ cells to organize cellular niches supporting negative selection of thymocytes [30].

References

1. Van Ewijk W (1991) T-cell differentiation is influenced by thymic microenvironments. Ann Rev Immunol 9:591-615
2. Forster R, Schubel A, Breitfeld D, Kremmer E, Renner-Muller I, Wolf E and Lipp M (1999) CCR7 coordinates the primary immune response by establishing functional microenvironments in secondary lymphoid organs. Cell 99:23-33
3. Dalmau SR, Freitas CS and Savino W (1999) Upregulated expression of fibronectin receptors underlines the adhesive capability of thymocytes to thymic epithelial cells during the early stages of differentiation: lessons from sublethally irradiated mice. Blood 93:974-990
4. Norment AM, Bogatzki LY, Gantner BN and Bevan MJ (2000) Murine CCR9, a Chemokine Receptor for Thymus-Expressed Chemokine That Is Up-Regulated Following Pre-TCR Signaling. J Immunol 164:639-648
5. Wurbel MA, Philippe JM, Nguyen C, Victorero G, Freeman T, Wooding P, Miazek A, Mattei MG, Malissen M, Jordan BR, Malissen B, Carrier A and Naquet P (2000) The chemokine TECK is expressed by thymic and intestinal epithelial cells and attracts double- and single-positive thymocytes expressing the TECK receptor CCR9. Eur J Immunol 30:262-271
6. Campbell JJ, Pan J and Butcher EC (1999) Cutting Edge: Developmental Switches in Chemokine Responses During T Cell Maturation. J Immunol 163:2353-2357
7. Burkly L, Hession C, Ogata L, Reilly C, Marconi LA, Olson D, Tizard R, Cate R and Lo D (1995) Expression of relB is required for the development of thymic medulla and dendritic cells. Nature 373:531-536

8. Naspetti M, Aurrand-Lions M, DeKoning J, Malissen M, Galland F, Lo D and Naquet P (1997) Thymocytes and RelB-dependent medullary epithelial cells provide growth-promoting and organization signals, respectively, to thymic medullary stromal cells. Eur J Immunol 27:1392-1397
9. Laufer TM, DeKoning J, Markowitz JS, Lo D and Glimcher LH (1996) Unopposed positive selection and autoreactivity in mice expressing class II MHC only on thymic cortex. Nature 383:81-85
10. Wu L, D'Amico A, Winkel KD, Suter M, Lo D and Shortman K (1998) RelB is essential for the development of myeloid-related CD8alpha- dendritic cells but not of lymphoid-related CD8alpha+ dendritic cells. Immunity 9:839-847
11. Malergue F, Galland F, Martin F, Mansuelle P, Aurrand-Lions M and Naquet P (1998) A novel immunoglobulin superfamily junctional molecule expressed by antigen presenting cells, endothelial cells and platelets. Mol Immunol 35:1111-1119
12. Martin-Padura I, Lostaglio S, Schneemann M, Williams L, Romano M, Fruscella P, Panzeri C, Stoppacciaro A, Ruco L, Villa A, Simmons D and Dejana E (1998) Junctional adhesion molecule, a novel member of the immunoglobulin superfamily that distributes at intercellular junctions and modulates monocyte transmigration. J Cell Biol 142:117-127
13. Aurrand-Lions M, Galland F, Bazin H, Zakharyev V, Imhof BA and Naquet P (1996) Vanin-1, a novel GPI-linked perivascular molecule involved in thymus homing. Immunity 5:391-405
14. Borkowski TA, Nelson AJ, Farr AG and Udey MC (1996) Expression of gp40, the murine homologue of human epithelial cell adhesion molecule (Ep-CAM), by murine dendritic cells. Eur J Immunol 26:110-114
15. Vremec D, Zorbas M, Scollay R, Saunders DJ, Ardavin CF, Wu L and Shortman K (1992) The surface phenotype of dendritic cells purified from mouse thymus and spleen: Investigation of the CD8 expression by a subpopulation of dendritic cells. J Exp Med 176:47-58
16. Nelson AJ, Dunn RJ, Peach R, Aruffo A and Farr AG (1996) The murine homolog of human Ep-CAM, a homotypic adhesion molecule, is expressed by thymocytes and thymic epithelial cells. Eur J Immunol 26:401-408
17. Simon B, Podolsky DK, Moldenhauer G, Isselbacher KJ, Gattoni-Celli S and Brand SJ (1990) Epithelial glycoprotein is a member of a family of epithelial cell surface antigens homologous to nidogen, a matrix adhesion protein. Proc Natl Acad Sci USA 87:2755-2759
18. Litvinov SV, Balzar M, Winter MJ, Bakker HA, Briaire-de Bruijn IH, Prins F, Fleuren GJ and Warnaar SO (1997) Epithelial cell adhesion molecule (Ep-CAM) modulates cell-cell interactions mediated by classic cadherins. J Cell Biol 139:1337-1348
19. Atlan-Gepner C, Naspetti M, Valéro R, Barad M, Lepault F, Vialettes B and Naquet P (2000) Disorganization of thymic medulla precedes evolution towards diabetes in female NOD mice. Autoimmunity 00:1-12
20. Savino W, Boitard C, Bach J-F and Dardenne M (1991) Studies on the thymus in nonobese diabetic mouse. I. Changes in the microenvironmental compartments. Lab Invest 64:405-417
21. Takeoka Y, Taguchi N, Kotzin BL, Bennett S, Vyse TJ, Boyd RL, Naiki M, Konishi J, Ansari AA, Shultz LD and Gershwin ME (1999) Thymic microenvironment and NZB mice: the abnormal thymic microenvironment of New Zealand mice correlates with immunopathology. Clin Immunol 90:388-398
22. Szala S, Froehlich M, Scollon M, Kasai Y, Steplewski Z, Koprowski H and Linnenbach AJ (1990) Molecular cloning of cDNA for the carcinoma-associated antigen GA733-2. Proc Natl Acd Sci USA 87:3542-3546
23. Perez MS and Walker LE (1989) Isolation and characterization of a cDNA encoding the KS1/4 epithelial carcinoma marker. J Immunol 142:3662-3667
24. Strnad J, Hamilton AE, Beavers LS, Gamboa GC, Apelgren LD, Taber LD, Sportsman JR, Bumol TF, Sharp JD and Gadski RA (1989) Molecular cloning and characterization of a human adenocarcinoma/epithelial cell surface antigen complementary DNA. Cancer Res 49:314-317
25. Bergsagel PL, Victor-Kobrin C, Timblin CR, Trepel J and Kuehl WM (1992) A murine cDNA encodes a pan-epithelial glycoprotein that is also expressed on plasma cells. J Immunol 148:590-596

26. Balzar M, Bakker HA, Briaire-de-Bruijn IH, Fleuren GJ, Warnaar SO and Litvinov SV (1998) Cytoplasmic tail regulates the intercellular adhesion function of the epithelial cell adhesion molecule. Mol Cell Biol 18:4833-4843
27. Litvinov SV, Velders MP, Bakker HA, Fleuren GJ and Warnaar SO (1994) Ep-CAM: a human epithelial antigen is a homophilic cell-cell adhesion molecule. J Cell Biol 125:437-446
28. Litvinov SV, van Driel W, van Rhijn CM, Bakker HA, van Krieken H, Fleuren GJ and Warnaar SO (1996) Expression of Ep-CAM in cervical squamous epithelia correlates with an increased proliferation and the disappearance of markers for terminal differentiation. Am J Path 148:865-875
29. Momburg F, Moldenhauer G, Hammerling GJ and Moller P (1987) Immunohistochemical study of the expression of a Mr 34,000 human epithelium-specific surface glycoprotein in normal and malignant tissues. Cancer Res 47:2883-2891
30. Naquet P, Naspetti M and Boyd R (1999) Development, organization and function of the thymic medulla in normal, immunodeficient or autoimmune mice. Sem Immunol 11:47-55

Genetic Dissection of Thymus Development

M. Schorpp[1], W. Wiest[1], C. Egger[1], M. Hammerschmidt[2], T. Schlake[1], and T. Boehm[1]

[1] Department of Developmental Immunology, [2] Spemann Laboratories, Max-Planck-Institute for Immunobiology, Stuebeweg 51, D-79108 Freiburg, Germany

Development of the Thymus

Lymphoid organs have developed to facilitate and to regulate the interaction of lymphocytes with non-lymphoid cells. In the mouse, pro-thymocytes originate in the fetal liver or the bone marrow; once they have entered the thymus, a complex series of differentiation steps leads to the generation of thymocytes bearing a diverse repertoire of T-cell receptors. The thymocytes are then subjected to various selection processes to eventually yield a population of T cells that exhibits self tolerance and restriction to self major histocompatibility complex. Because thymus development depends on reciprocal stromal-lymphoid interactions (1), it is disrupted by mutations affecting the differentiation of both the microenvironment (2) and T cells (3).

The alymphoid thymic anlage is the first morphologically defined step of thymus development (Fig. 1). The molecular mechanisms that lead to the formation of the alymphoid thymic anlage and the molecular characteristics that subsequently allow a productive homing by pro-thymocytes are unknown. Nevertheless, at some point during embryogenesis, epithelial cells in the third pharyngeal pouch receive a signal(s) that lays down the initial positional information for thymus development. Once this primordium of thymic microenvironment has been specified, it begins to develop via a series of reciprocal epithelial-mesenchymal interactions (1,2). One gene whose function may be important in this process is *Hoxa3*. *Hoxa3* is expressed in pharyngeal mesenchyme and pharyngeal endoderm and its inactivation

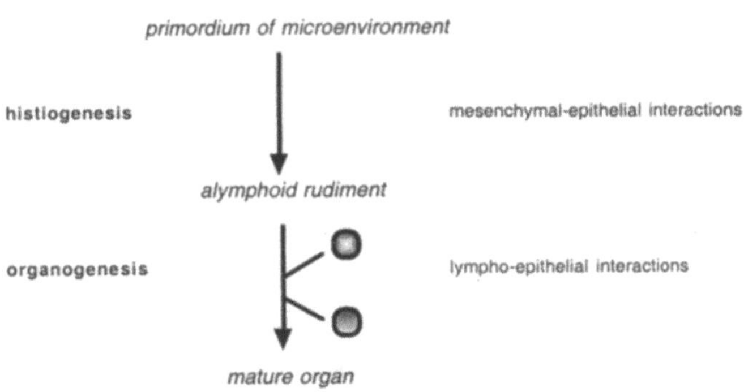

Fig. 1. Steps in the formation of lymphoid organs. Before entry of lympho-haematopoietic cells, an alymphoid rudiment forms from a primordium of the microenvironment. Organogenesis ensues with the immigration of lymphocytic progenitor cells.

interferes with the development of the alymphoid thymic anlage (4). *Pax9* is expressed in the pharyngeal endoderm and its targeted disruption likewise interferes with thymus development (5); it is not yet known at what stage the defect in thymus formation in *Pax9-/-* mice occurs. However, once the alymphoid thymic anlage has formed (Fig. 1), the function of the *Whn* (*nude*) gene (6) that is expressed in the third pharyngeal pouch endoderm is required for thymopoiesis, as *Whn-/-* (*nude*) mice have an arrested differentiation of the thymic epithelium (7). Even in the absence of Whn activity, the developing thymus moves caudally from a lateral position in the pharynx into a more ventral location of the superior mediastinum (Fig. 2).

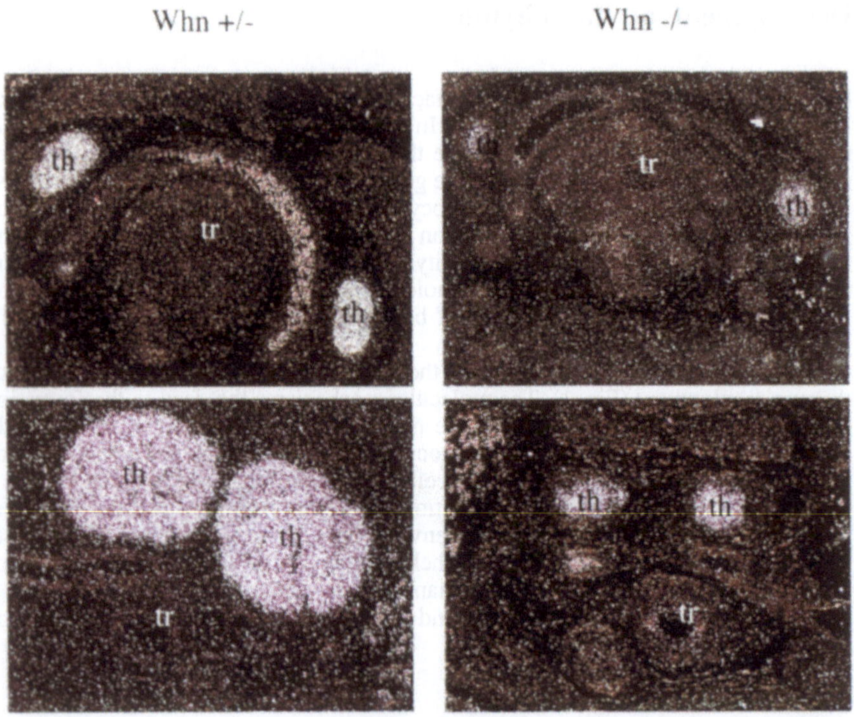

Fig. 2. Medio-caudal migration of the developing thymic lobes during mouse development. The thymic epithelium is identified via in situ hybridisation using a Whn anti-sense probe (white grains). The migration is not affected in mice with a disrupted *Whn* gene (Whn-/-). Note, however, that the size of the developing thymus is severely reduced in mutant mice. The wild-type thymus grows primarily due to the massive expansion of thymocytes. Top panels, dE12.5; bottom panels, dE14.5. Tr, trachea; Th, thymus.

The Function of the Whn Transcription Factor

The molecular characteristics of the thymic anlage that ensure the seeding by prothymocytes are unknown, so are the details of the ensuing reciprocal epithelial-lymphoid interactions. The *Whn* gene has been shown to be expressed in all epithelial cells of the thymus already before the entry of lymphoid progenitors (7) and thus is a tissue-specific marker of the thymic anlage. Mutations of the *Whn* gene are allelic to nude (7); to date, six spontaneously occurring nude alleles in human (8), mouse (6,9-12) and rat (6,9,13,14) have been studied at the molecular level (Fig. 3). All mutations lead to the characteristic nude phenotype of athymia and hairlessness. Two functionally important domains in the Whn transcription factor have been identified via sequence comparisons (13), and functional (10,13) analyses. The DNA binding domain of Whn proteins belongs to the so-called winged-helix/forkhead class (15) and represents an evolutionarily conserved branch of this class of proteins (16), as suggested by a detailed comparison of the closely related Whn and HTLF sub-families identified in non-vertebrates and vertebrates (data not shown). Although target genes of the Whn transcription factor in the thymic epithelium have not yet been identified, sequence-specificity of Whn´s DNA binding domain has been demonstrated (17). Indeed, a single mis-sense mutation of Whn encoded by the mouse *nu*Y allele eliminates functional activity of Whn (10). The transcriptional activation domain of Whn is of the acidic class and its absence inactivates Whn function (13).

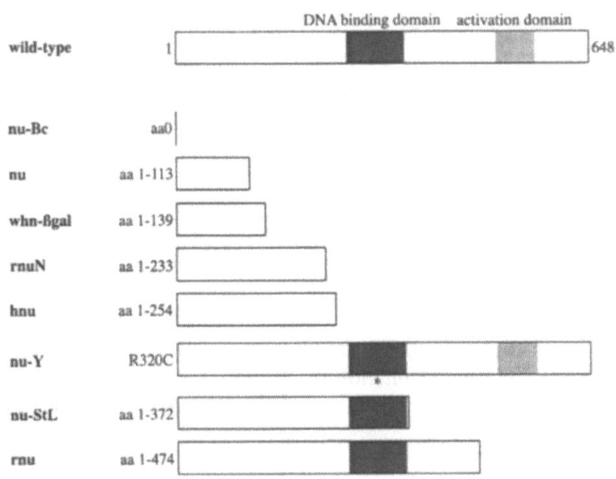

Fig. 3. Predicted structure of Whn proteins encoded by human, mouse and rat nude alleles. The structure of predicted mutant Whn proteins is shown beneath a schematic of the wild-type Whn protein. The different *nu* alleles are indicated on the left along with the length of the wild-type protein contained in these mutants. The molecular features of previously analysed nude alleles are as follows: *nu*Bc, aberrant splicing due to transposon insertion in intron upstream of first coding exon (11); *nu*, one bp deletion in exon 3 leading to translational frame-shift (6); *nu*ßgal, targeted disruption in exon 3; insertion of *IRES-ßgal* to convert *Whn* locus into bi-cistronic (7); *nu*rnuN, aberrant splicing due to transposon insertion in intron upstream of exon 5, which encodes N-terminal part of DNA binding domain (6,14); nuhnu, non-sense mutation in exon (8); *nu*Y, mis-sense mutation (R320C) in exon 7 leading to a non-functional Whn (10); *nu*StL, 2bp insertion in exon 7 leading to translational frame-shift (12); *nu*rnu, non-sense mutation in exon (9,13).

A Genetic Screen for Thymus Mutants in Zebrafish

To fully understand the genetic basis of lymphoid organ formation, all genes affecting the differentiation of T cell and microenvironment must be known. We have therefore embarked on a genetic analysis of thymus formation in zebrafish. Zebrafish are small tropical fish that reproduce rapidly and in great numbers. Several laboratories have reported the use of zebrafish to perform saturation mutagenesis for developmental mutants (for instance, 18,19). Zebrafish is an excellent model for vertebrate genetic analysis: eggs can be fertilized externally; embryos are completely transparent greatly facilitating morphological analysis; females produce large number of eggs accelerating genetic mapping.

We have used gynogenetic diploid embryos (20) produced from females heterozygous for ENU-induced mutations in a pilot screen to detect mutations affecting thymus development. Eggs fertilized with UV-inactivated sperm are subjected to high hydrostatic pressure within 6 min after fertilization, which suppresses the second meiotic division. A drawback of this so-called early pressure (EP) method is that the yield of mutant embryos depends on the chromosomal location of mutated loci (while mutations located close to centromers are detected in about 50% of EP-treated embryos, meiotic recombination and cross-over interference progressively reduce the percentage of mutant embryos for lesions located towards telomers); this is however off-set by the speed with which the screen can be performed.

In order to screen for aberrations in the formation of the thymus, in situ hybridization using a zebrafish *Rag-1* probe was employed, as the thymus is not morphologically detectable in living embryos. *Rag-1* is expressed in thymocytes after their entry into the thymic anlage (21). Using this method, we expect to be able to detect mutations affecting either the thymic microenvironment and/or the early stages of T cell development. We have found that using *Rag-1* as a probe, the thymus is first detectable 84 hours after fertilization (p.f.) as a bi-lateral structure emanating from the third pharyngeal pouch. The thymus then grows very rapidly and is clearly visible by day 5 p.f. Consequently, embryos are screened at 5 p.f. for size, shape and location of the thymus. Once presumptive mutant females are identified, they are used to establish F2 families for confirmation. About 25% of embryos should display the mutant phenotype in the F3 generation when male and female carriers are intercrossed, irrespective of the chromosomal location of the mutation.

Zebrafish Lines Carrying Mutations Affecting Thymus Development

Using the above described gynogenetic screen, we have established zebrafish lines which carry recessive mutations affecting thymus development. Extrapolating the numbers obtained in our current pilot screen, about 25-50 lines can be expected for an experiment approaching saturation of the zebrafish genome. To illustrate the features of mutant zebrafish, the characteristics of line 11-4 will be described. When embryos are allowed to develop at the normal temperature of 28.5°C and are tested for the presence of *Rag-1* expressing thymocytes in the thymus at 5 days p.f., the thymus is reduced to less than 50% of its normal size (Fig. 4A). When embryos are raised at slightly higher temperatures (31.5°C), no thymus develops as detected via

Rag-1 in situ hybridization (data not shown). This result indicates that T cell development is arrested at a very early stage. When +/? and -/- embryos are subjected to histological analysis, the thymus is indeed much smaller in mutant (left) as compared to normal (right) fish (Fig. 4B).

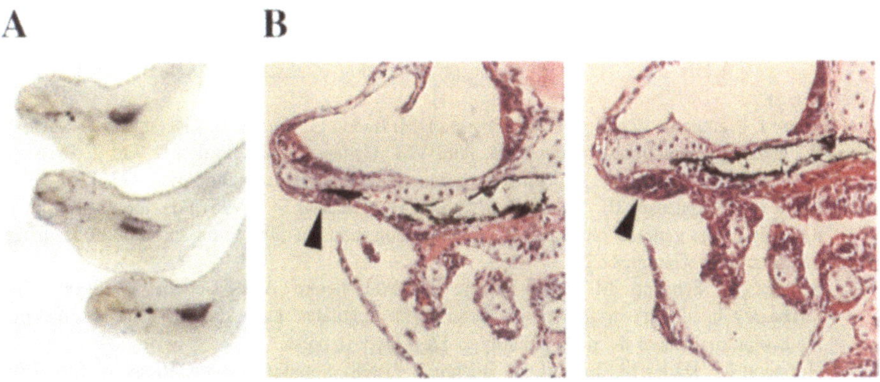

Fig. 4. Detection of thymus in zebrafish embryos via in situ hybridization with *Rag-1*. (**A**) Thymus malformation in zebrafish mutant 11-4. Note the smaller hybridization signal in one embryo (middle) of three F3 animals. (**B**) Histological analysis of wild-type (right panel) and mutant (left panel) embryos. Note the much reduced size of the mutant thymus. Sections were stained with haematoxylin-eosin; original magnification 200x.

Conclusion

Analysis of mice carrying specific mutations has resulted in the identification of genes that affect thymus development by interfering with differentiation of microenvironment and T cell lymphocytes. A systematic genetic analysis is required for a full understanding of the development of this primary lymphoid organ. We have decided to perform this mutational screen in zebrafish rather than mice for reasons of economy and speed. Nevertheless, since zebrafish genes are very similar to mouse homologs, it will be relatively easy to isolate appropriate mouse homologs and to use this information for site-directed mutagenesis in mice.

References

1. van Ewijk W, Shores EW, Singer A (1994) Cross-talk in the mouse thymus. Immunol Today 15: 214-217.
2. Boehm T, Nehls M, Kyewski B (1995) Transcription factors that control development of thymic microenvironment. Immunol Today 16: 555-556.
3. Kisielow P, von Boehmer H (1995) Development and selection of T cells: facts and puzzles. Adv Immunol 58: 87-209.

4. Manley NR, Capecchi M (1995) The role of Hoxa-3 in mouse thymus and thyroid development. Development 121: 1989-2003.
5. Peters H, Neubüser A, Kratochwil K, Balling R (1998) Pax9-deficient mice lack pharyngeal pouch derivatives and teeth and exhibit craniofacial and limb abnormalities. Genes Dev 12: 2735-2747.
6. Nehls M., Pfeifer D., Schorpp M., Hedrich H. Boehm T. (1994) New member of the winged-helix protein family disrupted in mouse and rat nude mutations. Nature 372: 103-107.
7. Nehls M Kyewski B Messerle M Waldschütz R Schüddekopf K Smith AJH, Boehm T (1996) Two genetically separable steps in the differentiation of thymic epithelium. Science 272: 886-889.
8. Frank J, Pignata C, Panteleyev AA, Prowse DM, Baden H, Weiner L, Gaetaniello L, Ahmad W, Pozzi N, Cserhalmi-Friedman PB, Aita VM, Uyttendaele H, Gordon D, Ott J, Brissette JL, Christiano AM (1999) Exposing the human nude phenotype. Nature 398: 473-474.
9. Segré JA, Nemhauser JL, Taylor BA, Nadeau JH, Lander ES (1995) Positional cloning of the nude locus: genetic, physical, and transcription maps of the region and mutations in the mouse and rat. Genomics 28: 549-559.
10. Schlake T, Schorpp M, Maul-Pavicic A, Malashenko AM, Boehm T (2000) The forkhead/winged-helix transcription factor Whn regulates hair keratin gene expression: molecular analysis of the nude phenotype. Dev. Dyn. in press.
11. Hofmann M, Harris M, Juriloff D, Boehm T (1998) Spontaneous mutations in SELH/Bc mice due to insertions of early transposons: molecular characterization of null alleles at the nude and albino loci. Genomics 52: 107-109.
12. Schorpp M, Schlake T, Kreamalmeyer D, Allen PM, Boehm T (2000) Molecular characterization of a novel nude allele, nu^{StL}, with antimorphic properties. submitted.
13. Schüddekopf K, Schorpp M, Boehm T (1996) The whn transcription factor encoded by the nude locus contains an evolutionarily conserved and functionally indispensable activation domain. Proc. Natl. Acad. Sci. USA 93: 9661-9664.
14. Huth M, Schlake T, Boehm T (1997) Transposon-induced splicing defect in the rat nude gene. Immunogenetics 45: 282-283.
15. Lai E, Clark KL, Burley SK, Darnell JE Jr (1993) Hepatocyte nuclear factor 3/fork head or "winged helix" proteins: a family of transcription factors of diverse biologic function. Proc. Natl. Acad. Sci. USA 90: 10421-10423.
16. Schorpp M, Hofmann M, Dear TN, Boehm T (1997) Characterization of mouse and human nude genes. Immunogenetics 46: 509-515.
17. Schlake T, Schorpp M, Nehls M, Boehm T (1997) The nude gene encodes a sequence-specific DNA binding protein with homologs in organisms that lack an anticipatory immune system. Proc. Natl. Acad. Sci. USA 94: 3842-3847.
18. Driever W, Solnica-Krezel L, Schier AF, Neuhauss SCF, Malicki J, Stemple DL, Stanier DYR, Zwartkruis F, Abdeliah S, Rangini Z, Belak J Boggs C (1996) A genetic screen for mutations affecting embryogenesis in zebrafish. Development 123: 37-46.
19. Haffter P, Granato M, Brand M, Mullins MC, Hammerschmidt M, Kane DA, Odenthal J, van Eeden FJM, Jiang Y-J, Heisenberg C-P, Kelsh RN, Furutani-Seiki M, Warga RM, Vogelsang E, Beuchle D, Schach U, Fabian C, Nüsslein-Volhard C (1996) The identification of genes with unique and essential functions in the development of the zebrafish, *Danio rerio*. Development 123:1-36.
20. Westerfield M (1994) The zebrafish book: a guide for the laboratory use of zebrafish. Eugene, Oregon, University of Oregon Press.
21. Willett CE, Zapata AG, Hopkins N, Steiner LA (1997) Expression of zebrafish rag genes during early development identifies the thymus. Dev Biol 182: 331-341.
22. Clark KL, Halay ED, Lai E, Burley SK (1993) Co-crystal structure of the HNF-3/fork head DNA-recognition motif resembles histone H5. Nature 364: 412-420.

Developing Thymocytes Organize Thymic Microenvironments

W. van Ewijk[1], H. Kawamoto[2], W. T.V. Germeraad[3], and Y. Katsura[2]
[1] Department of Immunology, Erasmus University Rotterdam, Rotterdam, The Netherlands
[2] Department of Immunology, Institute for Frontier Medical Sciences, Kyoto University, Kyoto, Japan
[3] Department of Immunology, University Hospital Utrecht, Utrecht, The Netherlands

A Functional Role of Lymphoid Microenvironments

The complex mechanisms underlying the development and functioning of the immune system depend on microenvironments in primary and secondary lymphoid organs as regulatory elements. Within a system of millions of different migratory cell types, microenvironments allow cells to meet and to interact with each other. Cellular interaction between specialized cell types and local variations in the concentration and type of soluble mediators control the divergence of immunological reactions within microenvironments.
Microenvironments are basically composed of a network of stromal cells in which other cell types derived from haematopoietic origin are integrated. In peripheral lymphoid organs, fibroblasts form the basic stromal network throughout the organ. Within this network, focal accumulations of interdigitating reticular cells (dendritic cells) regulate the homing of T lymphocytes in the T cell zones [1], while follicular dendritic cells determine the homing of B lymphocytes in follicles [2]. Macrophages are strategically located in domains where lymph is flowing [3]. They are not sessile, rather they migrate through the lymph node and present ingested antigen in microenvironments where antigen-reactive T and B cells are found in close proximity [4]. Upon entering lymphoid organs, B and T lymphocytes migrate through the complex organized lymphoid stroma and they home in their respective microenvironments. Addressins, homing receptors and chemokines are all mediators controlling the traffic pathways for lymphoid cells within lymphoid organs [5].

Likewise, within primary lymphoid organs, microenvironments control migration and development of hemopoietic progenitor cells. Here, differences in the composition of cell types within microenvironments crucially control the development of different hematopoietic lineages [6].

Both in the adult bone marrow, as well as in all peripheral lymphoid organs, fibroblasts are the basic structural cellular elements of the lymphohemopoietic stroma.

Thymic Microenvironments are Composed of Epithelial Cells

In striking contrast, in the thymus – the major primary organ for the development of T lymphocytes – epithelial cells form the basic elements within thymic microenvironments [7]. Apparently, developing T cells favour epithelial cells rather than fibroblasts. Fibroblasts do occur in the thymus, however they form a minority of stromal cells, and most of these cells are integrated in the capsule, trabeculae, peri-vascular spaces, and in the cortico-medullary junction [8].

Epithelial cells in the thymus differ in organization compared to epithelial cells in other (non-lymphoid) organs. In most organs, epithelial cells are organized in a two-dimensionally (2-D) oriented fashion. They are placed on a basement membrane, and form sheets of cells, separating the inside world from the outside world. Curiously, epithelial cells in the thymus have organized themselves in a three-dimensionally (3-D) organized network, similar to the network structure created by fibroblasts in the bone marrow and in peripheral lymphoid organs [9].

The network structure formed by epithelial cells clearly differs between the two major thymic compartments, the cortex and the medulla. In the cortex, epithelial cells form long cytoplasmic processes, interacting with each other through desmosomes. In the midcortex the cytoplasmic processes of epithelial cells are oriented more or less perpendicular to the thymic capsule. In this way, epithelial cells form structural elements in the migratory pathway of developing thymocytes on their way from the outer cortex to the medulla [7, 10]. Medullary epithelial cells do not show such long processes; rather, these cells are rounded and show secretory activity [7]. Dendritic cells and fibroblasts are mostly found at the cortico-medullary junction. Interestingly, thymic B lymphocytes also accumulate in this microenvironment.

Presently, the consensus is that the thymic stroma controls the proliferation, differentiation and selection of thymocytes [11-13]. Lymphostromal interaction continually occurs during a 3-week transit of the developing thymocytes throughout the thymic parenchyma. Thus, upon diapedesis through the venous vasculature at the cortico-medullary junction [14], T cell progenitors migrate through the thymic parenchyma to the outer subcapsular area. Here, developing T cells start proliferation and differentiation. Upon cell surface expression of T cell receptor (TCR) molecules, thymocytes start physical interactions with peptide-MHC complexes expressed by epithelial cells [7, 15]. Although the exact role of thymic microenvironments in selection is still unclear, the general idea is that, while migrating from the subcapsular cortex in the direction of the medulla, interaction of developing thymocytes with cortical epithelial cells and fibroblasts leads to positive selection [16], while interaction with dendritic cells and medullary epithelial cells induces tolerance to self epitopes [17]. The net balance between both types of lymphostromal interactions determines the daily production of thymocytes [18].

Thymic Crosstalk *in vivo*

Correct organization of thymic microenvironments is of prime importance for the development of the T cell repertoire. Over the past years it has become increasingly clear that developing thymocytes control the architecture of the thymic stroma, a process designated as thymic "crosstalk" [19]. In initial experiments, we found that the thymus of SCID mice has a well-developed 3-D organized epithelium in the cortex, but lacks medullae [20]. Transplantation of

SCID mice with bone marrow cells derived from normal mice restored (epithelial) microenvironments in the medulla, showing for the first time that thymic epithelial microenvironments can be remodelled by cells derived from hematopoietic origin. By crossing various types of TCR transgenic mice to SCID mice, the role of (mature) TCRαβ expressing thymocytes in the establishment of the epithelial microenvironments in the medulla was confirmed [21]. Similarly, infusion of SCID mice with mature peripheral $CD3^+$ T cells restored microenvironments in the thymic medulla [22].

Crosstalk also occurs in the thymic cortex. Mice expressing human CD3ε as a transgene show an arrest early in T cell development, at the level of DN-stage 1 ($CD44^+$, $CD25^-$) phenotype [23]. The thymic stroma in these transgenic mice is severely disturbed, and distinction between cortex and medulla is no longer possible. Moreover, the thymus of CD3ε transgenic mice showed several thymic cysts, lined by a 2-D organized epithelium consisting of cell types found in the respiratory and gastrointestinal tract. We have shown that this aberrant stromal phenotype can be repaired by infusion of T cell progenitors during embryogenesis, or in the first week of life [24,25]. Recently performed transplantation experiments using CD3ε transgenic mice, RAG^{null} mice and normal mice showed that progression of T cell development beyond the level of DN stage I thymocytes is of crucial importance to re-establish the 3-D network structure in the cortex of the thymus of these transgenic mice [25].

Thus, different types of developing thymocytes signal to different thymic stromal partners, to induce and maintain properly organized microenvironments. In turn, these microenvironments induce defined steps in the T cell development. This mutual interdependence between developing thymocytes and thymic epithelial cells, guarantees that thymopoiesis is a continuous flowing process, where developing thymocytes migrate through different microenvironments.

Thymic Crosstalk *in vitro*

Recently performed unpublished *in vitro* experiments have provided further insight in the mechanism of thymic "crosstalk" (Germeraad, Kawamoto, Katsura and van Ewijk, manuscript in preparation). The results of these experiments indicate that developing T cells already during embryogenesis critically regulate the 3-D organization of the thymic stroma. These experiments are based on the application of a fetal thymic organ culture system, where fetal thymic lobes are cultured submersed in tissue culture medium in the presence of a local high oxygen pressure (High Oxygen Submersion culture system; HOS i.e. 70% oxygen 5% CO_2 25% N_2; see also ref. 26). Thymocyte yield in this system is comparable to the "classical" fetal thymic organ culture system where fetal thymic lobes are cultured on a filter floating in tissue culture medium [26, 27].

In studying the architecture of microenvironments in fetal thymic lobes with monoclonal antibodies and single chain antibodies with specificity for subsets of epithelial cells [28], we have found that the thymic epithelial stroma during embryogenesis shows an impressive plasticity. Culture of Ed-14 fetal thymic lobes derived from normal mice under HOS condition shows, 6 days after onset of the culture, a well-developed 3-D organized epithelial reticulum in the peripheral

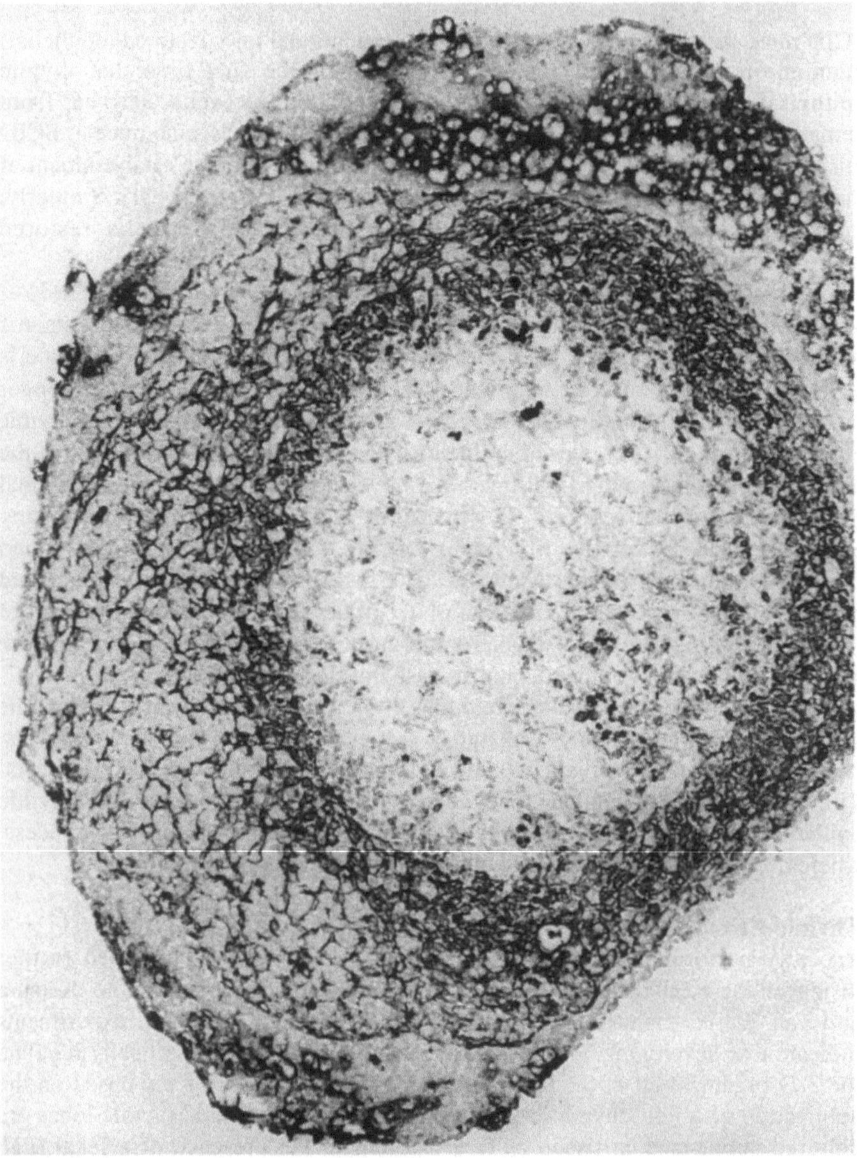

Fig. 1. Fetal thymic lobe, isolated from a Ed 14 $C_{57}Bl$ embryo, cultured for 6 days under High Oxygen Submersion (HOS) conditions. Frozen sections were labeled with TB4-4, a single chain antibody detecting cortical epithelial cells [28]. The peripheral part of this thymic lobe contains (unstained) fibroblasts and fat cells. The thymic epithelium shows a 3-D organization in the peripheral part of the lobe, and is more condensed in the mid part of the lobe. The central part of the lobe contains cell debris. (x 300).

part of the lobe (Fig. 1). In the mid part of the thymic lobes, epithelial cells form a ring of more closely packed cells, while in the center cell debris is usually present.

Surprisingly, fetal thymic lobes, both cultured in the HOS system, as well as in the classical filter system, do not develop extended medullae composed of ERTR5$^+$ epithelial cells (data not shown). Rather, ERTR5$^+$ cells remain dispersed throughout the lobe, and remain intermingled with ERTR4$^+$ cortical epithelial cells. Despite this lack of positioning of thymic medullae, all thymocyte subsets (CD4$^-$CD4$^-$; CD4$^+$ CD8$^+$; CD4$^+$ and CD8$^+$) develop normally [26].

The submersion culture system allows for the regulation of the oxygen concentration in the culture system. Lowering of the oxygen pressure (LOS condition; i.e. normal air, containing 5% CO_2) blocks T cell development at the

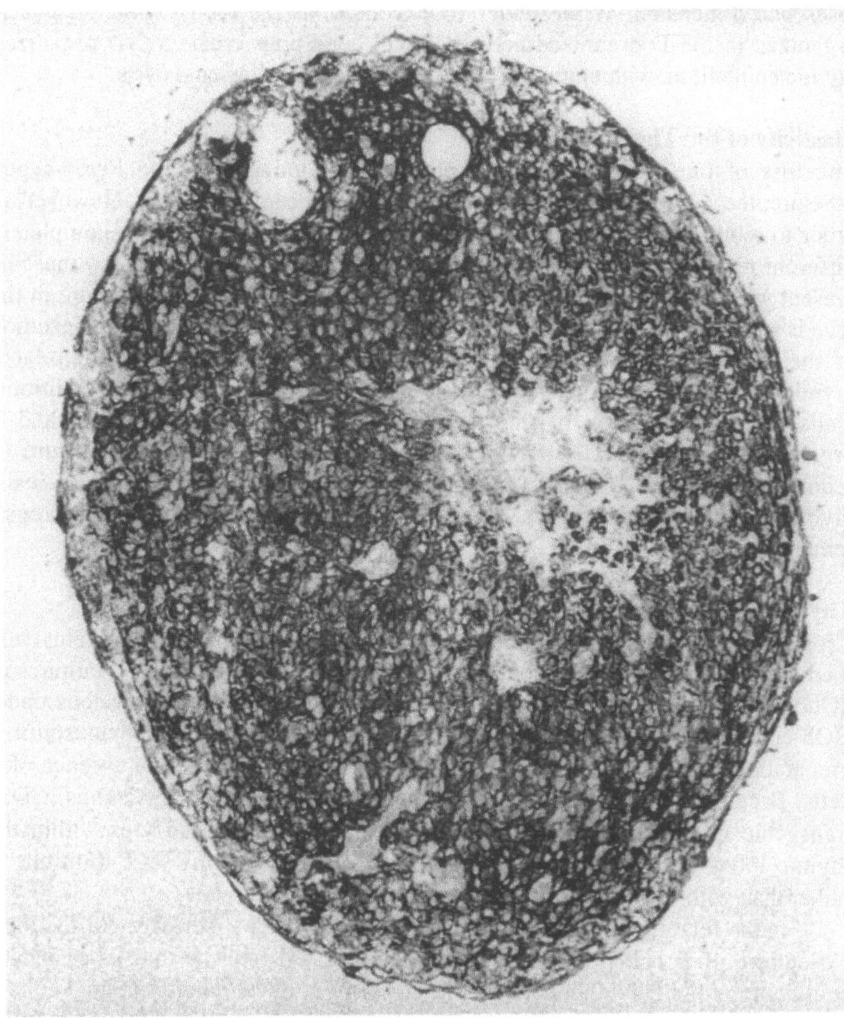

Fig. 2. Fetal thymic lobe obtained at Ed 14 cultured for 6 days under Low Oxygen Submersion (LOS) conditions. Frozen sections were labeled with TB4-4, a single chain antibody directed to cortical epithelial cells [28]. Epithelial cells are clustered and several thymic cysts are present. (x 300).

DN stage. Under these conditions lymphoid cells remain alive, but they are not able to proliferate and to differentiate further down the T cell lineage. Immunohistological analysis of the thymic microenvironments in lobes cultured under LOS conditions reveals a striking change in the organization of the epithelial stroma (Fig. 2). Both cortical and medullary epithelial cells withdraw their long cytoplasmic extensions within a culture period of 6 days, to form rounded cell clusters enveloped by a basement membrane. Moreover, several cysts appear, consisting of monolayers of epithelial cells underlined by a basement membrane.

In short, under LOS conditions, the organization of the thymic epithelium loses one dimension. While under HOS conditions the fetal thymic stroma is organized in a 3-D organized network, LOS conditions create a 2-D organized thymic epithelium, with epithelial cells positioned in clusters and cysts.

Plasticity of the Thymic Stroma

The loss of one dimension is not permanently induced by the low oxygen pressure; the 3-D organization of the thymic stroma can be restored. However, in order to rebuild the tertiary configuration of the thymic stroma, two completely different requirements must be fulfilled. First, developing thymocytes must be present within the thymic reticulum. Second, a local high oxygen pressure in the lobe is an absolute prerequisite to add the third dimension back in the organization of the epithelial stroma. Thus, co-culture of de-oxyguanosin (dGuo) treated thymic lobes with Sca-1^{lo} hematopoietic progenitor cells under HOS conditions, leads to full restoration of thymopoiesis and proper organization of a 3-D organized thymic epithelial reticulum within a period of 6 days. Culture of reconstituted lobes under LOS conditions does not restore thymopoiesis, thymocytes stay arrested at the DN stage, while epithelial cell clusters and cysts remain present.

Thymic Crosstalk *in vitro* is T Cell Specific

Crosstalk of developing thymocytes to thymic epithelial cells is exclusively mediated by developing thymocytes. We have recently performed co-cultures of dGuo treated fetal thymic lobes with other hematopoietic progenitor cells under HOS conditions (Germeraad, Kawamoto, Katsura, and Van Ewijk, manuscript in preparation). From *in vivo* experimental models it is known that in absence of T cells, B cells also can develop within thymic microenvironments. Thus, CD3ε transgenic mice contain a high frequency of IgM^+ B lymphocytes within the thymic lobes [29]. Such B cells mainly co-localize with $ERTR7^+$ fibroblasts, rather than with $ERTR4^+$ or $ERTR5^+$ epithelial cells [24].

In the fetal thymus, fibroblasts are mainly situated at periphery of the lobe. Co-culture of B cell progenitors with dGuo treated fetal thymic lobes indeed results in thymic B lymphopoiesis. The majority of developing lymphoid cells is $B220^+$, IgM^+ while a minority of B cells in the reconstituted lobes is IgG^+. Similarly to B cells in the thymus of CD3ε transgenic mice, B cells *in vitro* do not tend to grow within epithelial microenvironments and these cells are unable to restore the 3-D configuration of the thymic epithelium. Instead, a majority of B cells localizes within the peripheral rim of $ERTR7^+$ fibroblastoid cells.

Interestingly, the frequency of ER-TR7+ fibroblasts increases under these culture conditions, which indicates that developing B cells crosstalk to thymic fibroblasts, rather than to the epithelial cells.

Concluding Remarks

In summary, our *in vitro* data support the previously published *in vivo* data [19, 24] and show the impressive plasticity of the thymic epithelium during thymic embryogenesis. We find that the difference between a 2-D and a 3-D configuration of the thymic stroma is dependent on the presence of (1) developing thymocytes and (2) a high local oxygen pressure within the thymic lobes. Interestingly, B cells are able to differentiate along the B cell lineage within fetal thymic lobes. However, developing T and B cells crosstalk to different stromal partners: T cells prefer interaction with epithelial cells, whereas B cells interact with fibroblasts.

It is striking that thymopoiesis requires a high local oxygen pressure in the thymic lobes. The reason for this oxygen dependence is at present unknown. But it might explain why thymopoiesis is so efficiently performed within a primary lymphoid organ developing in an embryonic region which, in vertebrates, also supports the development of gills and the respiratory tract.

References

1. Ewijk W van, Verzijden JHM, Kwast ThH van der, Luijcx-Meijer SWM (1974) Reconstitution of the thymus dependent area in the spleen of lethally irradiated mice. Cell Tiss Res. 149:43-60.
2. Ewijk W van, Kwast ThH van der (1980) Migration of B lymphocytes in lymphoid organs of lethally irradiated, thymocyte-reconstituted mice. Cell Tissue Res. 212:497-508.
3. Ewijk W van, Brons NHC, Rozing J (1975) Scanning electron microscopy of homing and recirculating lymphocyte populations. Cell Immunol. 19:245-261.
4. Ewijk W van, Rozing J, Brons NHC, Klepper D (1977) Cellular events during the primary immune response in the spleen. A fluorescence, light and electronmicroscopic study in germfree mice. Cell Tissue Res. 183:471-489.
5. Butcher FC, Williams M, Youngman K. Rott L, Briskin M (1999) Lymphocyte trafficking and regional immunity. 72:209-253.
6. Kincade PW (1994) B lymphopoieses: globale factors, local control. Proc. Natl. Acad. Sci. 91:2888-2892.
7. Ewijk W van, Rouse RV, Weissman IL (1980) Distribution of H-2 microenvironments in the mouse thymus. Immunoelectron microscopic identification of I-A and H-2K bearing cells. J Histochem Cytochem. 28:1089-1099.
8. Vliet E van, Melis M, Ewijk W van (1984) Monoclonal antibodies to stromal cell types of the mouse thymus. Eur. J. Immunol. 14:524-529.
9. Ewijk W van (1988) Cell surface topography of thymic microenvironments. Lab. Invest. 59:579-590.
10. Weissman IL (1967) Thymus cell migration. J. Exp. Med. 126:291-295.

11. Boyd RL, Tucek CL, Godfrey ID, Izon DJ, Wilson TJ, Davidson NJ, Bean AGD, Ladyman HM, Ritter MA, Hugo P (1993) The thymic microenvironment. Immunol. Today 14:445-459.
12. Ewijk W van (1991) T-cell differentiation is influenced by thymic microenvironments. Annu. Rev. Immunol. 9:591-615.
13. Anderson G, Moore NC, Owen JJT, Jenkinson EJ (1996) Cellular interactions in the thymocyte development. Annu. Rev. Immunol. 14:73-99.
14. Ceredig R, Schreier M (1984) Immunohistochemical localization of host and donor derived cells in the regenerating thymus of radiation bone marrow chimeras. Thymus 6:15-26.
15. Jameson SC, Bevan MJ (1998) T cell selection. Curr. Opin. Immunol. 10:214-219.
16. Anderson G, Jenkinson EJ, Moore NC, Owen JJT. (1993) MHC class II positive epithelium and mesenchyme are both required for T cell development in the thymus. Nature 362:70-73.
17. Laufer T, Glimcher LH, Lo DL (1999) Using thymus anatomy to dissect T cell repertoire selection. Sem. Immunol. 11:65-70.
18. Anderson G, Partington KM, Jenkinson EJ (1998) Differential effects of peptide diversity and stromal cell type in positive and negative selection in the thymus. J. Immunol. 15:6599-6603.
19. Ewijk W van, Shores EW, Singer A (1994) Crosstalk in the mouse thymus. Immunol. Today 15:214-217.
20. Shores EW, Ewijk W van, Singer A (1992) Disorganization and restoration of thymic medullary epithelial cells in T cell receptor-negative Scid mice: evidence that receptor-bearing lymphocytes influence maturation of the thymic microenvironment. Eur. J. Immunol. 21:1657-1661.
21. Shores EW, Ewijk W van, Singer A (1994) Maturation of medullary thymic epithelium requires thymocytes expressing fully assembled CD3-TCR complexes. Int. Immunol. 6:393-1402.
22. Surh CD, Ernst B, Sprent J (1992) Growth of epithelial cells in the thymic medulla is under control of mature T cells. J. Exp. Med. 176:611-616.
23. Wang B, Biron C, She J, Higgins K, Sunshine MJ, Lacy E, Lonberg N, Terhorst C (1994) A block in both early T lymphocyte and natural killer cell development in transgenic mice with high copy numbers of the human CD3ε gene. Proc. Natl. Acad. Sci. USA 91:9402-9406.
24. Holländer GA, Wang B, Nichogiannopoulou A, Platenburg PP, Ewijk W van, Burakoff SJ, Gutierrez-Ramos J-C, Terhorst C (1995a) Developmental control point in induction of thymic cortex regulated by a subpopulation of prothymocytes. Nature 373:350-353.
25. Ewijk W van, Holländer G, Terhorst C, Wang B (2000) Stepwise development of thymic microenvironments in vivo is regulated by thymocyte subsets. Submitted for publication.
26. Watanabe Y, Katsura Y (1993) Development of T cell receptor αβ bearing T cells in the submersion organ culture of murine fetal thymus at high oxygen concentration. Eur. J. Immunol. 23:200-205.
27. Jenkinson EJ, Owen JJ (1990) T cell differentiation in thymus organ cultures. Semin. Immunol. 2:51-58.
28. Ewijk W van, De Kruif J, Germeraad WTV, Berendes P, Rökpe C, Platenburg PP, Logtenberg T (1997) Subtractive isolation of phage displayed single-chain antibodies to thymic stromal cells by using intact thymic fragments. Proc. Natl. Acad. Sci. USA 94:3903-3908.
29. Takoro Y, Sugawara T, Yaginuma H, Nakauchi H, Terhorst C, Wang B, Takahama Y (1998) A mouse carrying a genetic defect in the choice between T and B lymphocytes. J. Immunol. 161:4591-4598.

The Role of Mesenchyme in Thymus Development

J.J.T. Owen, D.E. McLoughlin, R.K. Suniara and E.J. Jenkinson
Department of Anatomy, Medical School, The University of Birmingham, Edgbaston, Birmingham B15 2TT, UK

Introduction: Thymic Mesenchyme is Derived from the Neural Crest

The embryological origins of the thymus are complex. The epithelial component arises from the third pharyngeal pouch and during its descent in the neck, it is surrounded by mesenchyme which is derived from the cephalic region of the neural crest [1]. Beginning at the 11th to 12th days of gestation in the mouse embryo, the thymus is colonised by haematopoietic stem cells which arrive via local blood vessels and then migrate through the mesenchyme into the epithelial rudiment. These stem cells give rise to lymphocytes, macrophages and dendritic cells [2].

It is known that the epithelial component of the thymus provides signals which are necessary for thymocyte differentiation. For example, epithelial cells are the source of the cytokine IL-7 [3] which is important for T-cell development [4] and, in addition, they present MHC/peptide complexes during the process of positive selection [5]. The role of mesenchyme cells is less well studied and in this contribution we will describe experiments which indicate that they are of importance especially during the early phases of T-cell development.

Mutations which Affect Neural Crest Migration and/or Function Influence Thymic Development

In humans, the DiGeorge syndrome results from a genetic defect which causes craniofacial abnormalities and limited thymic development. The cellular basis for this syndrome is thought to lie in a failure of neural crest migration. Bockman and Kirby [6] have produced a similar defect in chick embryos by removing the cephalic region of the neural crest prior to its migration to the pharyngeal region. In addition, targeted disruption of the rae 28 gene, which is thought to control Hox gene expression, results in defects in neural crest related tissues including the thymus [7]. Hoxa-3 gene disruption itself affects neural crest and thymic development [8]. In addition, these effects are accompanied by reduced expression of pax-1 in neural crest derived mesenchyme and, interestingly, both pax-1 and pax-3 mutant mice show thymic defects associated with abnormal neural crest function [9, 10].

In summary, these studies indicate that neural crest-derived mesenchyme cells are important for thymic development.

In vitro Studies on the Role of Mesenchyme in Thymic Development

Auerbach [11] was the first to show a role for mesenchyme in thymic development in vitro. He removed the peri-thymic mesenchyme from the 12-day mouse embryo thymus and noted a failure of lobulation of the latter during organ culture. However, culture conditions at the time were inadequate to determine whether lymphopoiesis was also affected. More recently, Shinohara and Honjo [12, 13] repeated these experiments and noted a failure of lobulation and MHC class I and class II expression in cultures of early 12-day embryo thymus following the removal of peri-thymic mesenchyme. They reported that epidermal growth factor could restore lobulation and insulin-like growth factors 1 and II could restore MHC class II but not class I expression. They did not study the effects of removal of mesenchyme on thymopoiesis. In other studies, Amagai et al [14] noted a failure of lymphopoiesis in organ cultures of 12-day mouse embryo thymus despite the fact that the thymus is known to be colonised by lymphoid stem cells at this stage. They did not state whether mesenchyme was removed from the 12-day thymic lobes prior to culture and so it is not clear whether the lower level of thymopoiesis was a consequence of the absence of mesenchyme. In a subsequent study [15] they reported that thymopoiesis could be restored by co-culture with fibroblastic cell lines.

We have repeated these experiments but in order to determine the role of mesenchyme, we have compared thymopoiesis in organ cultures of 12-day embryo thymus surrounded by mesenchyme as opposed to 12-day thymus lobes from which the peri-thymic mesenchyme was removed by micro-dissection [16]. The results are clear-cut and show that 12-day lobes consisting of epithelium and mesenchyme generate all thymus lymphocyte sub-populations including $CD4^+$ and $CD8^+$ T-cells, $\gamma\delta^+$ T-cells and NK cells. However in those lobes from which mesenchyme has been removed, there is a block in lymphoid development with only a small number of $CD4^-8^-$ cells remaining. Hence, these experiments show that peri-thymic mesenchyme does play an important role in T-cell development in vitro.

Is there an Indirect Effect of Mesenchyme on Thymopoiesis Operating through its Influence on Epithelial Morphogenesis and/or is there a Direct Effect on Lymphocyte Precursors?

The experiments of Auerbach [11] and Shinohara and Honjo [12,13] on the failure of thymic lobulation in the absence of mesenchyme were interpreted as suggesting an effect of mesenchyme on epithelial morphogenesis. However thymic lobulation is dependent on the ingrowth of connective tissue septae containing fibroblasts derived from peri-thymic mesenchyme. Hence the failure of lobulation might simply be due to the absence of mesenchymal cells which are required for the formation of connective tissue septae. The results of mesenchymal removal on MHC expression are more difficult to interpret especially as the authors did not state whether their cultures underwent lymphoid development. Our own studies show that the removal of mesenchyme from 12-day lobes has a dramatic effect on lymphopoiesis and for the following reasons we suggest that this might be due to a direct effect of mesenchymal cells and fibroblasts on thymic lymphocytes:

(a) Using a cell marker approach we have found that cells of the peri-thymic mesenchyme migrate into the thymus shortly after the 12th day of gestation [16]. These cells not only form the connective tissue septae of the thymus but also form a network of cells closely associated with developing lymphocytes.

(b) We have shown previously [17] that thymic fibroblasts are necessary for the development of $CD4^-8^-$ thymocytes even when "mature" epithelial cells are used in reaggregate organ cultures.

(c) We have also shown that thymic fibroblasts are associated with an intra-thymic network of extracellular matrix (ECM) which is in direct contact with thymocytes and thus has the potential to influence the development of the latter [17].

Thymic Fibroblasts and the ECM: A Role in Thymopoiesis?

Savino et al [18] first demonstrated the presence of an extensive ECM within the thymus and suggested that it might play an important role in T-cell development. The assembly of ECM is known it be an active cellular process involving integrin activation as well as cytoskeletal activity [19]. The higher order structure of the matrix, namely the presence of repeated units of fibronectin as well as associated molecules such as laminin and proteoglycans is known to be important in cell activation [20]. In addition thymocytes express integrins capable of interacting with the ECM [21] and their development also depends upon the cytokine IL-7 which is known to bind to heparin sulphate proteoglycans and hence might be presented to developing T-cells by the ECM [22]. Thus the role of fibroblasts might depend on their ability to produce an ECM.

These possibilities require further investigation and as a first step, we are currently attempting to block ECM assembly in vitro in order to examine its effects on thymopoiesis.

Summary

We have reviewed the evidence that thymic mesenchymal cells and their progeny thymic fibroblasts play an important role in early T-cell development. Although it is possible that mesenchyme plays an inductive role in thymic epithelial morphogenesis, we have presented evidence to suggest that there is a direct effect of mesenchyme and fibroblasts on T-cell development. Moreover the association of these cell types with an ECM raises the possibility that the latter might be important in integrin and/or cytokine presentation especially during the $CD4^{-}8^{-}$ phase of T-cell development.

References

1. Le Lievre, C.S., and Le Douarin, N.M. (1975) Mesenchymal derivatives of the neural crest: analysis of chimeric quail and chick embryos. J. Embryol. Exp. Morphol. 34:125-154.
2. Anderson, G., Moore, N.C., Owen, J.J.T. and Jenkinson, E.J. (1996) Cellular interactions in thymocyte development. Annu. Rev. Immunol. 14:73-99.
3. Moore, N., Anderson, G., Smith, C.A., Owen, J.J.T. and Jenkinson, E.J. (1993) Analyses of cytokine gene expression in subpopulations of isolate thymocytes and thymic stromal cells using semiquantitative polymerase chain reaction. Eur. J. Immunol. 23:922-924.
4. Von Freeden-Jeffry, U., Vierra, P., Lucian, L.A., Mc Neill, T., Burdach, S.E. and Murray, R. (1995) Lymphopoiesis in interleukin (IL)-7 gene-deleted mice identifies IL-7 as a non redundant cytokine. J. Exp. Med. 181:1519-1526.
5. Anderson, G., Owen, J.J.T., Moore, N.C. and Jenkinson, E.J. (1994) Thymic epithelial cells provide unique signals for positive selection of $CD4^{+}CD8^{+}$ thymocytes in vitro. J. Exp. Med. 179:2027-2031.
6. Bockman, D.E. and Kirby, M.L. (1984) Dependence of thymic development on derivatives of the neural crest. Science. 223:498-500.
7. Takihara, Y., Tomotsune, D., Shirai, M., Katoh-Fukui, Y., Nishii, K., Motaleb, M.A,. Nomura, M., Tsuchiya, R., Fujita, Y., Shibata, Y., Higashinakagawi, T. and Shimada, K. (1997) Targeted disruption of the mouse homologue of the Drosophila polyhomeotic gene leads to altered anteroposterior patterning and neural crest defects. Development. 124:3673-3682.
8. Manley, N.R. and Capecchi, M.R. (1995) The role of Hoxa-3 in mouse thymus and thyroid development. Development. 121:1989-2003.
9. Dietrich, S. and Gruss, P. (1995) Pleiotropic undulated phenotypes suggest a role for Pax-1 for growth processes in mesoderm, endoderm and neural crest derived structures. Dev. Biol. 167:529-548.
10. Conway, S.J., Henderson, D.J. and Copp, A.J. (1997) Pax 3 is required for cardiac neural crest migration in the mouse: evidence from the splotch (Sp2H) mutant. Development. 124:505-514.
11. Auerbach, R. (1960) Morphogenetic interactions in the development of the mouse thymus gland. Dev. Biol. 2:271-285.
12. Shinohara, T. and Honjo, T. (1996) Epidermal growth factors can replace thymic mesenchyme in induction of embryonic thymus macrogenesis in vitro. Eur.J.Immunol. 26:747-752.
13. Shinohara, T. and Honjo, T. (1997) Studies in vitro on the mechanism of the epithelial/mesenchyme interaction in the early fetal thymus. Eur. J. Immunol. 27:522-529.
14. Amagai, T., Itoi, M. and Kondo, Y. (1995) Limited development capacity of the earliest embryonic murine thymus. Eur. J. Immunol. 25:757-762.

15. Itoi, M. and Amagai, T. (1998) Inductive role of fibroblast cell lines in development of the mouse thymus anlage in organ culture. Cell. Immunol. 183:32-41.
16. Suniara, R.K., Jenkinson, E.J. and Owen, J.J.T. (2000) An essential role for thymic mesenchyme in early T cell development. J Exp.Med. 191:1051-1056.
17. Anderson, G., Anderson, K.L., Tchilian, E.Z., Owen, J.J.T. and Jenkinson, E.J. (1997) Fibroblast dependency during early thymocyte development maps to the $CD25^{+}CD44^{+}$ stage and involves interactions with fibroblast matrix molecules. Eur. J. Immunol. 27:1200-1206.
18. Savino, W., Villa-Verde, D.M.S. and Lannes-Vieira, J. (1993) Extracellular matrix proteins in intrathymic T cell migration and differentiation. Immunol. Today. 14:158-161.
19. Wu, C., Keivans, V.M., O'Toole, T.E., McDonald, J.A. and Ginsberg, M.H. (1995) Integrin activation and cytoskeletal interaction are essential for the assembly of a fibronectin matrix. Cell. 83:715-724.
20. Streuli, C. (1999) Extracellular matrix remodelling and cellular differentiation. Current Opinion Cell Biol. 11:634-640.
21. Wadsworth, S., Halvorson, M.J., Chang, A.C. and Coligan, J.E. (1993) Multiple changes in VLA protein glycosylation, expression, and function occur during mouse T cell ontogeny. J. Immunol. 150:847-857.
22. Clark, D., Katoh, O., Gibbs, R.V., Griffiths, S.D. and Gordon, M.Y. (1995) Interaction of IL-7 with glycosaminoglycans and its biological relevance. Cytokine. 7:325-327.

Making Central T-Cell Tolerance Efficient: Thymic Stromal Cells Sample Distinct Self-Antigen Pools

B. Kyewski[1], B. Röttinger[1], and L. Klein[2]
[1]Tumor Immunology Program, German Cancer Research Center, INF 280, D-69120 Heidelberg, FRG
[2]Department of Cancer Immunology & AIDS, Dana Farber Cancer Institute, 44 Binney St., Boston, MA 02115, USA

Introduction

Induction and maintenance of self-tolerance in the developing and mature T-cell repertoire is mediated by multiple mechanisms operating both in the thymus ("central tolerance") and in peripheral lymphoid and non-lymphoid organs ("peripheral tolerance") [1,2]. The thymus is viewed as the prime site of T-cell tolerance induction to ubiquitous (e.g. house keeping genes or MHC antigens) and abundant blood-borne antigens entering the thymus via the circulation (e.g. C5 complement protein) [3-5]. Though intensely investigated, tolerance to self-antigens which are confined to specific tissues and organs often expressed by only a minor set of cells (e.g. ß-cells of the pancreas), has largely remained puzzling [6-8]. The prevailing view that tolerance induction via negative selection in the thymus applies only to a limited set of "abundant" proteins, however, has been challenged by recent findings that intrathymic expression of "tissue-specific" antigens is a common occurrence. This phenomenon is termed "ectopic" or "promiscuous" gene expression. Interestingly, this ectopic expression appears to be a unique property of thymic epithelial rather than bone marrow-derived antigen presenting cells [9-17]. Here we provide evidence that confinement of this interesting property to a certain stromal cell type, in conjunction with different efficiencies to present exogenous antigens, results in sampling by distinct thymic stromal cells of largely non-overlapping self-antigen pools. We will argue that this subtle and intricate functional complementation in tolerance induction by epithelial cells and hemopoietic cells extends the scope of central T-cell tolerance to a wide range of tissue-specific self-antigens.

Soluble antigens which oscillate between non-tolerogenic and tolerogenic levels present a particular challenge for the induction and maintenance of self-tolerance. Intrathymic tolerization of the continuously developing T-cell repertoire requires continuous presentation of self-antigens. However, in the case of inducibly secreted proteins with serum levels that can vary over a wide range, the blood-derived antigen supply may drop below a certain threshold that would allow antigen-specific high-avidity T-cells to exit the thymus. Thus temporal limitations of intrathymic antigen supply should prevent continuous central tolerization of CD4 T-cells. We recently developed a transgenic mouse model to study T-cell tolerance to such a candidate antigen, namely the major acute phase protein, human C-reactive protein (hCRP). Human C-reactive protein was expressed in transgenic mice under its autologous regulatory regions. Basal serum levels of hCRP in these mice are below 10^{-9} M and reach 5×10^{-7} M after experimental induction. Surprisingly, thymocytes expressing a transgenic T-cell receptor (TCR-tg) specific for a dominant epitope of hCRP were deleted at the $CD4^+8^+$ stage even at extremely low basal serum levels. It turned out that this efficient deletion was due to ectopic expression of hCRP by medullary epithelial cells, thus rendering induction of central tolerance independent of liver-derived protein [18]. Importantly, intrathymic expression of transgenic hCRP reflects a physiological property of human and mouse acute phase proteins.

Quantitative studies on the selection of the T-cell repertoire assigned a prominent role in negative selection to hemopoietic cells, in particular dendritic cells [19]. By contrast, intrathymic antigen presentation by (medullary) epithelial cells has been reported to result rather in induction of anergy [20-23]. In the hCRP model, expression is confined to medullary epithelial cells, yet the mechanism of tolerance is deletion. Hence, we addressed the question whether deletion of hCRP-specific T-cells by antigen expressed in medullary epithelial cells involves antigen transfer to dendritic cells as a mandatory step [24, 25], or whether medullary epithelial cells autonomously express and present antigen and mediate negative selection.

Results and Discussion

Medullary Epithelial Cells Autonomously Tolerize for Their Endogenous Antigens

We tested these alternatives by constructing mice in which medullary epithelial cells were the sole source of hCRP and at the same time presentation of hCRP by MHC-class-II-positive bone marrow-derived APC (i.e. dendritic cells, macrophages, and B cells) was abrogated. In brief, thymectomized BL/6 mice were transplanted with hCRP-transgenic fetal thymus grafts, lethally irradiated and reconstituted with TCR-tg bone marrow derived from mice which were either MHC-class-II-positive or deficient in MHC-class-II expression. Development of hCRP-specific, TCR-tg positive thymocytes in thymic grafts was analyzed by four color fluorescence [18]. As reported previously, hCRP expression in thymic epithelial cells is sufficient and necessary for deletion of hCRP-specific, TCR-tg thymocytes, when liver-derived hCRP was expressed at non-induced levels [18]. Importantly, the same degree of intrathymic deletion was observed when potential presentation of hCRP by hemopoietic APC was abrogated (Table 1). Thus transfer of hCRP to non-epithelial cells is not necessary for efficient deletion of a quasi-monoclonal T-cell repertoire. By implication, medullary epithelial cells serve an autonomous role in tolerance induction to hCRP, i.e. they express and present hCRP and induce deletion [26].

Thymic DC, but not Medullary Epithelial Cells Tolerize for Blood-Derived Self-Antigens

In a second set of mice, liver-derived, blood-borne hCRP was the only source of antigen and we assessed in the same way the respective role of epithelial cells and hemopoietic APC in mediating deletion of hCRP-specific thymocytes. Briefly, thymectomized hCRP-transgenic male BL/6 mice were transplanted with fetal wildtype BL/6 thymic grafts, lethally irradiated and reconstituted with TCR-tg bone marrow derived from MHC-class-II-positive or MHC-class-II-deficient mice. In contrast to hCRP expressed by medullary epithelial cells, in the case of circulating hCRP, deletion of specific thymocytes crucially depended on the presence of MHC-class-II-positive hemopoietic APC (Table 1). Strikingly, MHC-class-II-positive epithelial cells in these mice did not substitute for hemopoietic cells. It is concluded from this result that hemopoietic cells, most likely dendritic cells, are necessary for tolerance induction to high levels of circulating hCRP (note that male hCRP transgenic mice displaying constitutively induced levels of hCRP [18] were used in this setting). In addition, this finding implies that epithelial cells - cortical or medullary - do not capture, process and present exogenous hCRP sufficiently, even at rather high serum levels of 5×10^{-7} M, so as to mediate detectable deletion.

Table 1. Deletion of antigen-specific thymocytes by antigen expressed in medullary epithelial cells

intrathymic source of hCRP	wild-type	MHC-class-II-k.o.
	donor bone marrow cells	
mEpc*	+++**	+++
blood-borne	+++	— (+)

*mEpc: medullary epithelial cells
**The relative degree of deletion, as measured by the frequency of TCR-tg, CD4 single positive thymocytes, is indicated.
For details see text.

As implied by this type of study, in which the fate of hCRP-specific thymocytes was measured, epithelial cells and dendritic cells differ in their efficacy of sampling exogenous, blood-borne antigens. In a further experiment, we addressed this supposition directly. Balb/c mice were injected i.v. with varying doses of sperm whale myoglobin and 2 h later, thymic dendritic cells, macrophages, and cortical and medullary epithelial cells were purified as described [17] and co-cultured with myoglobin-specific T-T hybridomas. Activation of T-T hybridomas, as measured by interleukin-2 release, was analysed. The highest protein dose tested (0.5 mg protein/g body weight) led to presentation by dendritic cells and to a 10-fold lesser degree by medullary epithelial cells, and macrophages, whereas cortical epithelial cells remained non-stimulatory. When a 3 to 10-fold lower protein dose was injected, only dendritic cells, but neither epithelial cells nor macrophages would activate the T-T hybridomas (Table 2).

Table 2. Dose-dependent presentation of blood-borne exogenous antigen

Myoglobin (i.v.)	0.5	0.15 [mg/g body weight]	0.05
DC*	+++**	++	+
MΦ	++	—	—
mEpc	++	(+)	—
cEpc	—	—	—

* DC: dendritic cells; MΦ: macrophages; mEpc: medullary epithelial cells; cEpc: cortical epithelial cells.
** The relative degree of antigen-specific presentation is indicated.
Each increment corresponds approximately to a 10-fold difference.
For details see text.

These results clearly demonstrate a hierarchy in the capacity of thymic stromal cells to sample and present exogenous proteins entering the thymus via the circulation. Interestingly, even at the highest dose of protein pulse cortical epithelial cells were non-stimulatory revealing a further difference between epithelial cells in the cortex and the medulla. The inability of cortical epithelial cells to present myoglobin under these experimental conditions may be due to a very low rate of fluid phase uptake thus preventing protein capture within the short window of protein exposure [27, 28]. Alternatively, the blood-thymus barrier largely secludes the cortex from circulating proteins [29].

Complementation of Stromal Cell Subsets in Tolerance Induction

Taken together, these data suggest a functional distinction between medullary epithelial cells and dendritic cells with regard to MHC-class-II antigen processing pathways and tolerance induction: while epithelial cells efficiently present endogenous self-antigens including ectopically expressed tissue-specific proteins [9-17], dendritic cells preferentially present exogenous (circulating) self antigens. These differences result in the sampling by medullary stromal cells of largely non-overlapping self-antigen pools and the display of self-peptide profiles specific for each cell type (Fig. 1).

A particular feature of expression of peripheral antigens in the thymic medulla is its confinement to very few scattered cells. Analysis of hCRP expression by *in situ* hybridization revealed 1-3 positive cells per section preferentially located in the outer medulla. Similar findings have been reported for insulin, somatostatin, S100β, class 1b HLA-G, and beta-galactosidase driven by the promotor of the acetylcholine receptor [13, 30-33]. In the case of hCRP, this implies that few scattered medullary epithelial cells expressing this antigen autonomously mediate negative selection of a quasi-monoclonal T-cell repertoire. This striking efficacy can be best explained by extensive scanning of stromal cells in the medulla by developing thymocytes [34]. In conjunction with such serial cell-cell interactions between thymocytes and stromal cells ("serial scanning model"), differences in self-antigen display among medullary epithelial cells and dendritic cells would allow for maximal (re)presentation of the "hemopoietic and non-hemopoietic self" in the thymus and thus maximize the scope and efficacy of central tolerance [35, 36].

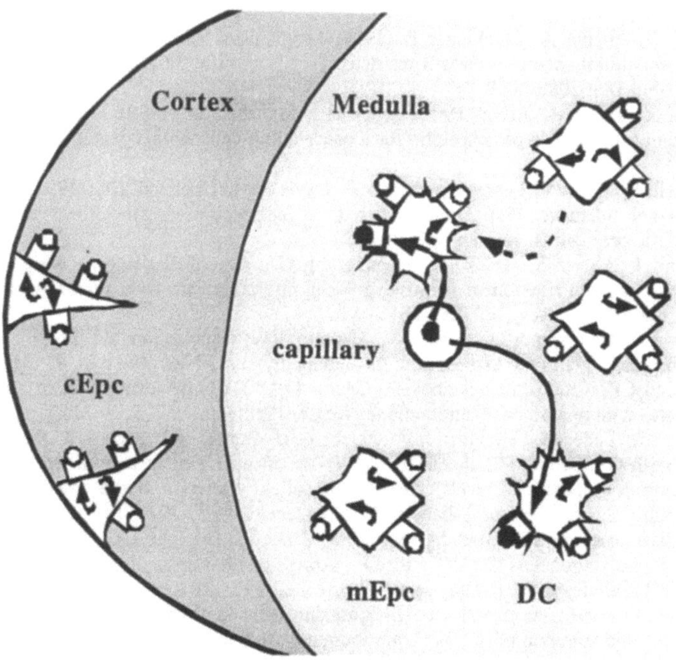

Fig. 1. Thymic stromal cells sample different self-antigen pools
While cEpc and mEpc cells preferentially present endogenous self-antigens via MHC-class-I and -II, DC are highly efficient in processing and presentation of blood-borne, extra-cellular self-antigens. The self-peptide profile of cEpc and mEpc cells thus largely reflects their respective genetic programs, while the self-antigen profile of thymic DC to a great extent represents the extracellular milieu. The model proposes that each cEpc and each DC expresses a similar self-peptide profile, whereas individual mEpc differ in their self-peptide profile. This diversification of self-antigen display among different stromal cell types maximizes the (re)-presentation of "self" in the thymus.
DC: dendritic cells; cEpc: cortical epithelial cells; mEpc: medullary epithelial cells.

Acknowledgements

We are most grateful to Sonja Höflinger and Klaus Hexel for expert technical assistance. We wish to acknowledge support by the Deutsche Forschungsgemeinschaft (grants Ky 7/6-1 and 7/8-1), the German Cancer Research Center (Heidelberg, FRG) and the International Agency for Research on Cancer (Lyon, France) (Research Training Fellowship to L.K.).

References

1. Kisielow P, von Boehmer H (1995) Development and selection of T-cells: facts and puzzles. Adv Immunol 58: 87-209
2. Stockinger B (1999) T lymphocyte tolerance: from thymic deletion to peripheral control mechanisms. Adv Immunol 71: 229-265
3. Kyewski BA, Fathman, CG, Kaplan, HS (1984) Intrathymic presentation of circulating non-major histocompatibility complex antigens. Nature 308: 196-199
4. Kappler JW, Roehm N, Marrack P (1987) T-cell tolerance by clonal elimination in the thymus. Cell 49: 273-280
5. Zal T, Volkmann A, Stockinger B (1994) Mechanisms of tolerance induction in major histocompatibility complex class II-restricted T-cells specific for a blood-borne self-antigen. J Exp Med 180: 2089-2099
6. Förster I, Hirose R, Arbeit JM, Clausen BE, Hanahan D (1995) Limited capacity for tolerization of CD4 T-cells specific for a pancreatic β-cell neo-antigen. Immunity 2: 573-585
7. Akkaraju S, Ho WY, Leong D, Canaan K, Davis MM, Goodnow CC (1997) A range of CD4 T-cell tolerance: Partial inactivation to organ-specific antigen allows nondestructive thyroiditis or insulitis. Immunity 7: 255-271
8. Alferink J, Aigner S, Reibke R, Hämmerling GJ, Arnold B (1999) Peripheral T-cell tolerance: the contribution of permissive T-cell migration into parenchymal tissues in the neonate. Immunol. Rev. 169: 255-261
9. Pribyl TM, Campagnoni C, Kampf K, Handley VW, Campagnoni AT (1996) The major myelin protein genes are expressed in the human thymus. J Neurosci. Res 45: 812-819
10. Egwuagu CE, Charukamnoetkanok P, Gery I (1997) Thymic expression of autoantigens correlates with resistance to autoimmune disease. J Immunol 159: 3109-3112
11. Pugliese A, Zeller M, Fernandez Jr A, Zalcberg LJ, Bartlett RJ, Ricordi J, Pietrapaolo M, Eisenbarth GS, Bennet ST, Patel DD (1997) The insulin gene is transcribed in the human thymus and transcription levels correlate with allelic variation at the *INS* VNTR-*IDDM2* susceptibility locus for type 1 diabetes. Nature Genet 15: 293-297
12. Vafiadis P, Bennet ST, Todd JA, Nadeau J, Grabs R, Goodyer CG, Wickramasinghe S, Colle E, Polychronakos C (1997) Insulin expression in human thymus is modulated by *INS* VNTR alleles at the *IDDM2* locus. Nature Genet 15: 289-292
13. Hanahan D (1998) Peripheral-antigen-expressing cells in thymic medulla: factors in self-tolerance and autoimmunity. Curr Opin Immunol 10: 656-662
14. Heath VL, Moore NC, Parnell SM, Mason DW (1998) Intrathymic expression of genes involved in organ-specific autoimmune diseases. J Autoimmun 11: 309-318
15. Werdelin O, Cordes U, Jensen T (1998) Aberrant expression of tissue-specific proteins in the thymus: a hypothesis for the development of central tolerance. Scand J Immunol 47: 95-100
16. Sospedra M, Ferrer-Francesch X, Dominguez O, Juan M, Foz-Sala M, Pujol-Borrell R (1999) Transcription of a broad range of self-antigens in human thymus suggests a role for central mechanisms in tolerance toward peripheral antigens J Immunol 161: 5918-5929

17. Klein L, Klugmann M, Nave K-A, Tuohy V K, Kyewski B (2000) Shaping of the autoreactive T-cell repertoire by a splice variant of self protein expressed in thymic epithelial cells. Nature Med 6: 56-61
18. Klein L, Klein T, Rüther U, Kyewski B (1998) CD4 T-cell tolerance to human C-reactive protein, an inducible serum protein, is mediated by medullary thymic epithelium. J Exp Med 188: 5-16
19. van Meerwijk JP, Marguerat S, Lees RK, Germain RN, Fowlkes BJ, MacDonald HR (1997) Quantitative impact of thymic clonal deletion on the T-cell repertoire. J Exp Med 185: 377-383
20. Hoffmann MW, Allison J, Miller JFAP (1992). Tolerance induction by thymic medullary epithelium. Proc Nat Acad Sci USA 89: 2526-2530
21. Schönrich G, Momburg F, Hämmerling GJ, Arnold B (1992) Anergy induced by thymic medullary epithelium. Eur J Immunol 22: 1687-1691
22. Antonia SJ, Geiger T, Miller J, Flavell RA (1994). Mechanisms of immune tolerance induction through the thymic expression of a peripheral tissue-specific protein. Int. Immunol. 7: 715-725.
23. Degermann S, Surh CD, Glimcher H, Sprent J (1994) B7 Expression on thymic medullary epithelium correlates with epithelium mediated deletion of $V\beta5^+$ thymocytes. J Immunol 152: 3254-3263
24. Humblet C, Rudensky A, Kyewski B (1994) Presentation and intercellular transfer of self-antigen within the thymic microenvironment: expression of the E alpha peptide-I-A^b complex by isolated thymic stromal cells. Int Immunol 6: 1949-1958
25. Viret C, Barlow AK, Janeway CA Jr (1999) On the intrathymic intercellular transfer of self-determinants. Immunol Today 20: 8-10
26. Oukka M, Colucci-Guyon E, Tran PL, Cohen-Tannoudji M, Babinet C, Lotteau V, Kosmatopoulos K (1996) CD4 T-cell tolerance to nuclear proteins induced by medullary thymic epithelium. Immunity 4: 545-553
27. Oukka M, Andre P, Turmel P, Besnard N, Angevin V, Karlsson L, Trans PL, Charron D, Bihain B, Kosmatopoulos K, Lotteau V (1997) Selectivity of the major histocompatibility complex class II presentation pathway of cortical thymic epithelial cell lines. Eur J Immunol 27: 855-859
28. Kasai M, Kominami E, Mizuochi T (1998) The antigen presentation pathway in medullary thymic epithelial cells, but not that in cortical thymic epithelial cells, conforms to the endocytic pathway. Eur J Immunol 28: 1867-1876
29. Nieuwenhuis P, Stet RJ, Wagenaar JP, Wubbena AS, Kampinga J, Karrenbeld A (1988) The transcapsular route: a new way for (self-) antigens to by-pass the blood-thymus barrier? Immunol Today 9: 372-375
30. Kojima K, Reindl M, Lassmann H, Wekerle H (1997) The thymus and self tolerance: Co-existence of encephalitogenic S100β-specific T cells and their nominal autoantigen in the normal adult rat thymus. Int Immunol 9: 897-904
31. Smith KM, Olson DC, Hirose R, Hanahan D (1997) Pancreatic gene expression in rare cells of the thymic medulla: Evidence for functional contribution to T cell tolerance. Int Immunol 9: 1355-1365
32. Crisa L, McMaster MT, Ishii JK, Fisher SJ, Salomon D (1997) Identification of a thymic epithelial cell subset sharing expression of the class 1b HLA-G molecule with fetal trophoblasts. J Exp Med 186: 289-298
33. Salmon AM, Bruand C, Cardona A, Changeux JP, Berrih-Aknin S (1998) An acetylcholine receptor alpha subunit promoter confers intrathymic expression in transgenic mice. Implications for tolerance of a transgenic self-antigen and for autoreactivity in myasthenia gravis. J Clin Invest 101: 2340-2350
34. Merkenschlager M, Power MO, Pircher H, Fisher AG (1999) Intrathymic deletion of MHC class I-restricted cytotoxic T-cell precursors by constitutive cross-presentation of exogenous antigen. Eur J Immunol 29: 1477-1486
35. Farr AG, Rudensky A (1998) Medullary thymic epithelium: a mosaic of epithelial 'self'? J Exp Med 188: 1-4
36. Klein L, Kyewski, B (2000) Self-antigen presentation by thymic stromal cells: a subtle division of labor. Curr Opin Immunol 12:179-186

IV

Lymphoid Tissue Development

T Cell Activation and Polarization by DC1 and DC2

Y-J. Liu, N. Kadowaki, M-C. Rissoan, and V. Soumelis
DNAX Research Institute, 901 California Avenue, Palo Alto, CA 94304 USA

Introduction

Dendritic cells (DCs) are professional antigen-presenting cells, capable of activating naïve T helper (TH) cells [1-3]. Differentiation of activated TH cells into IFN-γ producing effector TH1 cells or IL-4, -5 and -10 producing effector TH2 cells depends respectively on cytokines, such as IL-12 or IL-4, possibly produced by a third cell type [3-7] (Fig. 1). IL-12 produced by activated macrophages was believed to be critical for TH1 differentiation [5-7]. Because IL-4 is a TH2 prototype cytokine, the original source of IL-4 required for TH2 differentiation has been controversial [4, 5]. During early cognate DC-T cell interaction, activated T cells rapidly express T cell activation antigen, CD40-ligand [8-10] (Fig. 2). The finding that CD40-ligand and microbial product rapidly induce DCs to produce a large amount of IL-12 suggests that dendritic cells may not need a third cell type to polarize activated T cells towards TH1 effectors [11-14]. The questions are: i) are there distinct types of DCs? ii) Do distinct types of DCs induce different types of immune responses, such as TH1 versus TH2 or immunity versus tolerance? iii) how do cytokine microenvironment or innate immunity determine the functions of DCs? iv) is there an IL-4 independent mechanism for the induction of TH2 differentiation? v) what are the ideal DCs for tumor therapy or for treatment of autoimmune diseases and graft versus host diseases (GVHD)?

Lymphoid Versus Myeloid DCs in Mice

A lymphoid and a myeloid DC developmental pathway have been proposed in mice. Thymic DCs were derived from thymic lymphoid progenitors, which also have T and NK cell differentiation potential [15]. Myeloid DCs, together with macrophages and granulocytes were shown to be developed within the same colony from a single bone marrow progenitor cell cultured with GM-CSF [16]. Myeloid DCs were also shown to be differentiated from blood monocytes with GM-CSF and IL-4 [17]. In mouse spleen, whereas the $CD11c^{+}CD11b^{+}CD8^{-}$ marginal zone DCs were suggested to be derived from the myeloid pathway, $CD11c^{+}CD11b^{-}CD8^{+}$ DCs in the T cell areas were suggested to be derived from the lymphoid pathway [18]. There is a population of dendritic cells localized in the T cell areas of spleen and lymph nodes which expresses an intracellular antigen MIDC8 and no or low CD11c [19, 20]. This population of DCs appears to be derived from monocytes [20] and may have a relatively longer life span (unpublished observation).

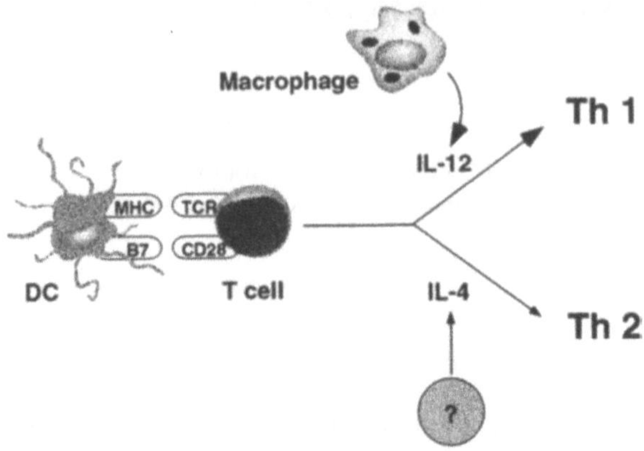

Fig. 1. Regulation of TH differentiation. DCs activate naïve TH cells due to their high expression of MHC class II and costimulatory molecules. Differentiation of activated TH cells into TH1 or TH2 depends respectively on IL-12 produced by activated macrophages or IL-4 produced by a third cell type such as NK T cells or eosinophils [4, 5].

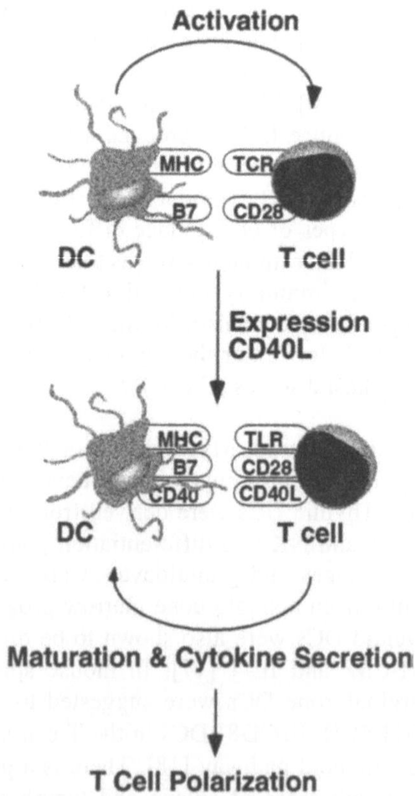

Fig. 2. Reciprocal interaction between DCs and T helper cells. TH cells rapidly express CD40-ligand after activation by DCs. CD40-ligand not only induces TH cells to express high MHC class II and costimulatory molecules such as CD80 and CD86, but also stimulates TH cells to secrete T cell-polarizing cytokines.

DC1 and DC2 in Human

The most direct evidence that humans may also have different DCs comes from the finding that human blood contains two distinct types of dendritic cell precursors. Monocytes (pre-DC1) were shown to differentiate into immature myeloid DCs (im-DC1) after 7-10 days of culture with GM-CSF and IL-4 [21, 22]. A second type of DC precursor (pre-DC2), previously known as plasmacytoid T cells [23] or plasmacytoid monocytes [24] was shown to differentiate into im-DC2 after 3-6 days of culture with IL-3 or monocyte conditional medium [25-34]. Unlike CD11c$^+$ immature DCs from blood and tonsils, both pre-DC1 and pre-DC2 were unable to differentiate into DCs in the absence of cytokines. Pre-DC1 and pre-DC2 display many different features: i) pre-DC1 but not pre-DC2 express MHC class-like molecules CD1a, b, c, and d,

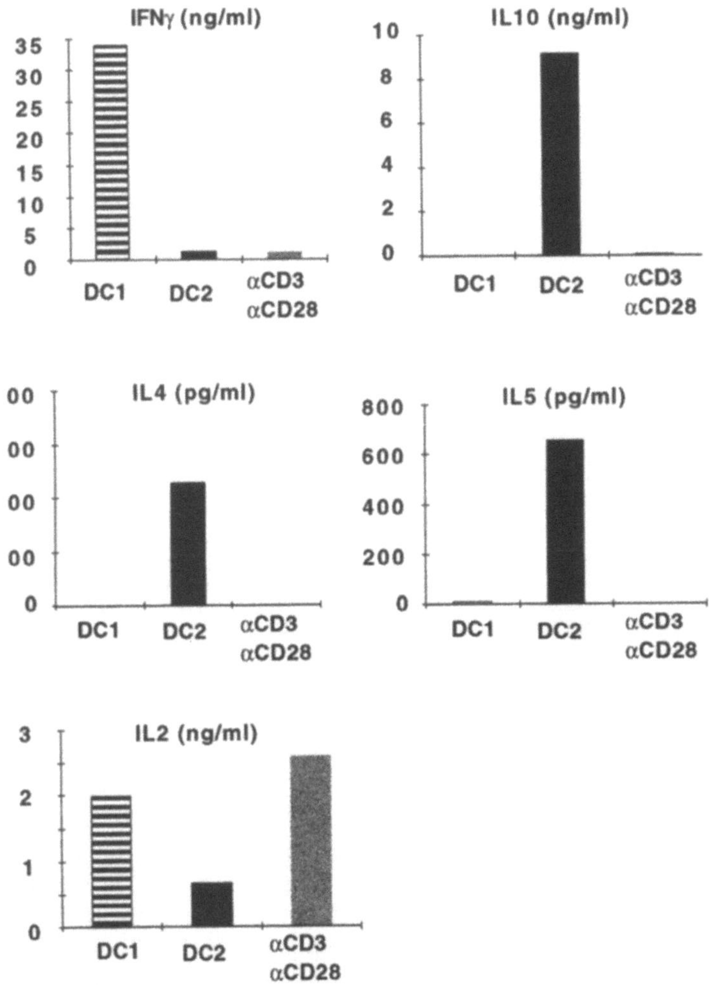

Fig. 3. Human CD4$^+$CD45RO$^-$ naïve T cells were cultured for 6 days with allogeneic CD40-ligand-activated DC1 or DC2 or anti-CD3 plus anti-CD28. Cells were counted and restimulated with anti-CD3 and anti-CD28 for 24 hours. Levels of IFN-γ, IL-4, IL-5, IL-10 and IL-2 within culture supernatants were collected after 24 hours and measured by ELISA.

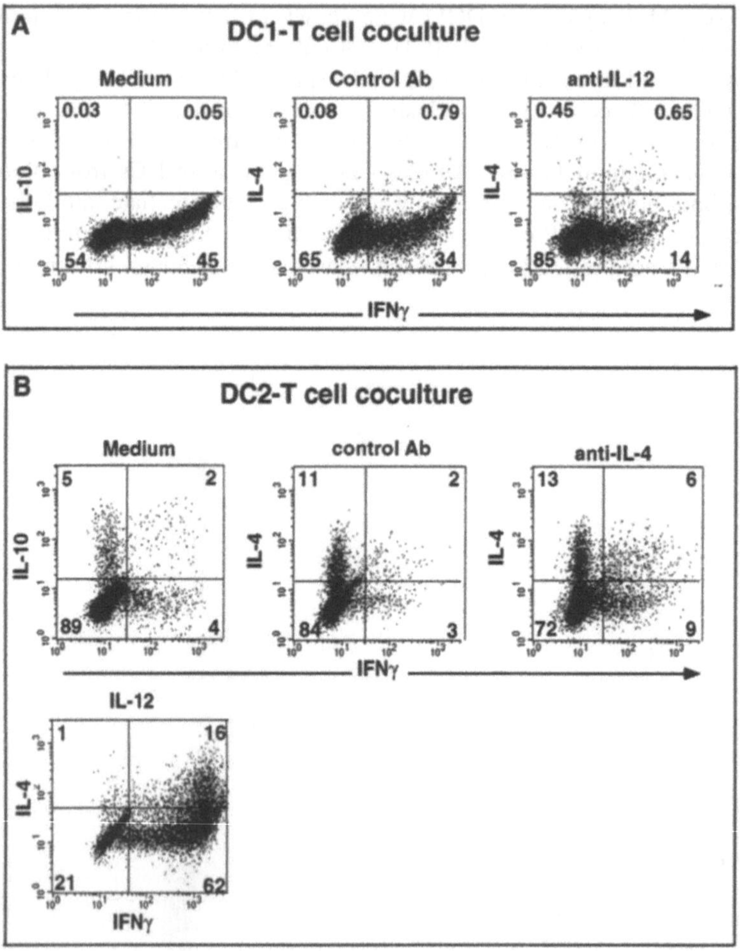

Fig. 4. Intracellular cytokine expression. A: DC1-T cell cocultures with control goat IgG antibodies and goat antibody to IL-12. B: DC2-T cell coculture with medium, goat Ig, goat antibody to IL-4 and IL-12. Double staining with anti-IL-4 plus anti-IFN-γ or anti-IL-10 plus IFN-γ was performed. 10^4 cells were analyzed and the percentage of each T cell population is indicated in the plots.

and mannose receptors, suggesting that pre-DC1 and pre-DC2 may recognize different types of antigens [34]. ii) while pre-DC1 depend on GM-CSF to differentiate into im-DC1, pre-DC2s mainly depend on IL-3 to differentiate into immature im-DC2. This correlates with high GM-CSFRα and low IL-3Rα on pre-DC1 and low GM-CSFRα and high IL-3Rα on pre-DC2 [25-27]. iii) While IL-4 is required for pre-DC1 to differentiate into im-DC1, IL-4 kills pre-DC2 when cultured with IL-3, indicating different regulations for the two DC differentiation pathways [25, 30]. iv) in contrast to pre-DC1 which express many myeloid antigens such as CD11b, CD11c, CD13, CD33 and mannose receptors, and are able to differentiate into macrophages when cultured with M-CSF, pre-DC2 express low or no myeloid antigens and are unable to differentiate into macrophages [25, 26]. v) pre-DC2 express high levels of pre-T cell receptor alpha chain (pre-Tα) suggesting that pre-DC2 may be differentiated from lymphoid precursors [33].

While im-DC1 Produce a Large Amount of IL-12, im-DC2 Produce a Low Amount of IL-12 after CD40-ligand Activation [25]

CD40-ligand not only triggers DC maturation [35, 36], but also stimulates DCs to produce cytokines [11-14]. While im-DC1 produced a large amount of IL-12p75 (over 500 pg/ml), im-DC2 produce a much lower amount of IL-12p75 (less than 50pg/ml) during 24 hours of CD40-ligand activation. This was confirmed by quantitative RT-PCR analyses showing that im-DC1 expressed 20 times more IL-12p40 mRNA (10,000-40,000 fg/50ng DNA) than do im-DC2 (500-600 fg/50ng cDNA) after CD40-ligand activation. In addition, CD40-ligand activated im-DC1 but not im-DC2 produced a significant amount of IL-1α, IL-1β, IL-6 and IL-10. Both CD40-ligand-activated im-DC1 and im-DC2 produce a comparable amount of chemokine IL-8. Neither DC1 nor DC2 produces IL-4 mRNA or protein.

CD40-ligand-activated DC1 and DC2 Respectively Induce TH1 Versus TH2 Differentiation [25]

Both CD40-ligand-activated DC1 and DC2 induce strong proliferation of naïve $CD4^+$ T helper cells. To examine the function of DC1 and DC2 in T helper cell differentiation, $CD4^+CD45RO^-CD45RA^+$ naïve T cells were cultured for 6 days with: i) 24 hour CD40-ligand activated DC1; ii) 24 hour CD40-ligand-activated DC2 or iii) anti-CD3 and anti-CD28 coated on a culture plate. After 6 days of priming, T cells were re-stimulated with anti-CD3 and anti-CD28 coated on a culture plate for either 5 hours to examine intracellular cytokine expression using flow cytometry or for 24 hours to examine cytokines secreted into the culture supernatants using ELISA. CD40-ligand-activated DC1 induced allogeneic naïve $CD4^+$ T cells to produce a large amount of IFN-γ (over 30,000pg/ml) and little IL-4, IL-5 and IL-10. By contrast, CD40-ligand-activated DC2 induced $CD4^+$ naïve T cells to produce a much lower level of IFN-γ (less than 500pg/ml), but significant levels of IL-4 (over 200pg/ml), IL-5 (over 600pg/ml), and IL-10 (over 800pg/ml) (Figs. 3 and 4). These results suggest that: i) DCs not only activate naïve T cells, but also directly polarize activated T cells toward TH1 and TH2 differentiation; ii) while CD40-ligand-activated DC1 induce TH1 differentiation, CD40-ligand-activated DC2 preferentially induce TH2 differentiation.

The fact that DC1-induced TH1 differentiation was only partially inhibited by anti-IL-12 antibody suggests that either IFN-γ or other undefined factors might be involved [25] (Fig. 4). The findings that DC2 do not produce IL-4 and that anti-IL-4 do not block DC2-induced TH2 differentiation suggest that DC2 may express an undefined TH2 differentiation factor. Trans-well culture experiments suggest that DC2-induced TH2 differentiation requires direct DC-T cell contact. In addition, exogenous IL-12 did not block DC2-induced generation of IL-4-producing cells (Fig. 4).

Similarly in mice, while CD11c+CD11b-CD8+ DCs were found to produce a high level of IL-12 in response to CD40-ligand and microbial stimulation and induce Th1 differentiation, CD11c+CD11b+CD8- DCs were found to produce a low level of IL-12 and induce TH2 differentiation [14, 37, 38].

Signals from Pathogens and Cytokines may Determine the Effector Function of ImDCs

ImDC1s derived from monocytes after 5-7 days of culture with GM-CSF and IL-4 have the potential to induce both TH1 and TH2 differentiation [39]. Two groups of signals were shown to stimulate imDC1 to induce Th1 differentiation: i) LPS derived from gram negative bacteria, gram positive bacteria particle SAC [40], unmethylated bacterial CpG-containing oligonucleoside [41-44] and double stranded viral RNA [45]; ii) T cell signals such as CD40-ligand [11-14] and IFN-γ [40]. Many signals were shown to stimulate imDC1 to induce TH2 differentiation or to inhibit TH1 differentiation: i) anti-inflammatory molecules such as IL-10, TGF-β, PGE2 and steroid [40, 46] and ii) OX40-ligand expressed on T cells [47, 48]. IFN-α/β may directly stimulate TH1 differentiation by activation of Stat 4 [49]. However, IFN-α/β may also inhibit TH1 differentiation through downregulating IL-12 production by DCs [50-53]. The clinical application of IFN-β in treating multiple sclerosis suggests that IFN-α/β may have anti-inflammatory functions. Some studies found that CD40-ligand strongly stimulates imDC1 to produce IL-12 and induce TH1 differentiation [11-14, 25]. Other studies found that CD40-ligand is insufficient to stimulate imDC1 to produce IL-12 to induce TH1 and exogenous IL-12 was required for CD40-ligand activated imDC1 to induce TH1 differentiation [39]. Two technical differences may need to be considered: i) soluble CD40-ligand and CD40-ligand-transfected L cells may have different potency in stimulating imDC1 to produce IL-12; ii) restimulation of primed T cells with anti-CD3 plus anti-CD28 or with PMA and ionomycine may have different results.

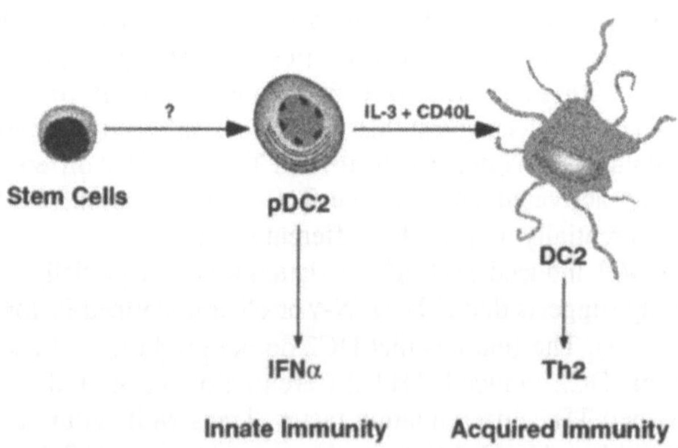

Fig. 5. Pre-CD2 link innate and adaptive immunity. Pre-DC2 rapidly produce a huge amount of type 1 IFN in response to viral invasion. Pre-DC2 also have the potential to differentiate into dendritic cells, capable of stimulating naïve T cells.

Dendritic Cells from Different Tissue Origin Display Different Direction in Polarizing T Helper Cell Differentiation

$CD11c^+$ DCs isolated from mouse mucosa such as Peyer's patches or respiratory tracts preferentially induce TH2 differentiation. By contrast, $CD11c^+$ DCs isolated from mouse spleen preferentially induce Th1 differentiation [54, 55]. DCs derived from bone marrow precursors or from liver precursors also displayed different effector function in polarizing T helper cells. While liver-derived DCs produced high IL-10 and induce allogeneic T cells to undergo TH2 differentiation, bone marrow-derived DCs produced low IL-10 and induce allogeneic T cells to undergo TH1 differentiation during primary MLR in cultures or in allogeneic recipient mice after DC transfer [56]. The functional differences among different tissue DCs may result from differences in tissue cytokine microenvironments as well as in the lineage origin of different tissue DCs.

Pre-DC2 are Natural Interferon-α/β Producing Cells

Pre-DC2, previously know as plasmacytoid T cells/monocytes [23, 24] display striking ultra-structural features that resemble immunoglobulin-secreting plasma cells: the juxtanuclear Golgi apparatus and parallel arrays of rough endoplasmic reticulum [26]. This cellular ultrastructure suggests that pre-DC2 may make and secrete large amounts of proteins and may therefore have important functions. Interestingly, pre-DC2 had many similar features to the elusive IFN-α-producing cells (IPCs) described by many groups [57-65]: i) low frequency in blood (about 0.1%); ii) expressing CD4 and MHC class II and lacking lineage markers; iii) die rapidly in cultures. Since these cells can be enriched together with DCs from blood, some studies concluded that IPCs were DCs which display dendritic morphology and are capable of inducing T cell activation [64]. By contrast, other studies suggested that IPCs were unable to induce T cell activation [62]. Two groups have recently demonstrated that pre-DC2s are identical to IPCs, which produce up to 1000 times more IFN than other blood cell types [66, 67]. Pre-DC2s may represent a unique cell lineage, which plays critical roles in both anti-viral innate immune responses and in triggering T cell-mediated adaptive immune responses (Fig. 5).

Conclusions and Future Perspectives

Dendritic cells play two critical roles in T cell-mediated immune responses: i) activate naïve T helper cells; and ii) polarize activated T helper cells towards TH1 or TH2 effector cells. During early cognate DC-naïve T cell interaction, T cells rapidly express CD40-ligand, which strongly stimulates DCs to upregulate MHC class I and II, CD80 and CD86, and to express/secrete T cell-polarizing molecules/cytokines. While CD40-ligand-activated DC1 produce a large amount of IL-12 and cause TH1 differentiation, CD40-ligand-activated DC2 produce a low level of IL-12 and mainly cause TH2 differentiation in an IL-4-independent fashion. T helper cell differentiation may depend on both the types of antigen-presenting cells and microenvironment signals that act on imDCs. From our studies, CD40-ligand-activated DC1 may be the most efficient natural adjuvant in vaccination against tumors and infectious pathogens, because DC1 likely induce strong inflammatory TH1 responses as well as strong CTL responses in vivo. By contrast, CD40-ligand-activated DC2 may have the potential to be used in treating TH1-mediated autoimmune diseases and GVHD, because DC2 may induce anti-inflammatory TH2 immune responses in vivo.

The demonstration of pre-DC2 as the professional natural IFN-α producing cells (IPCs) has ended the long search for the functional identity of the elusive IFN-α producing cells as well as the enigmatic plasmacytoid T cells/monocytes. Pre-DC2/IPCs may represent a novel effector cell type for anti-viral innate immunity, as well as a distinct type of DC precursors, which are able to initiate T cell-mediated adaptive immunity. These two functions of the pre-DC2/IPCs lineage may provide a cellular mechanism by which innate immunity connects with and shapes adaptive immunity (Fig. 5).

Acknowledgments

I would like to thank the members of my lab B. Blom, H. Kanzler, M. Gilliet, S. Ho and S. Antonenko for their contributions and critical reading of the manuscript. J. Katheiser for excellent editorial help.

References

1. Banchereau, J Steinman RM (1998) Dendritic cells and the control of immunity. Nature 392:245-252.
2. Cella M, Sallusto F, Lanzavecchia A (1997) Origin, maturation and antigen presenting function of dendritic cells. Curr Opin Immunol 9:10-16
3. Reis e Sousa C, Sher A, Kaye P (1999) The role of dendritic cells in the induction and regulation of immunity to microbial infection. Curr Opin Immunol 11:392-399
4. Abbas AK, Murphy KM, Sher A (1996) Functional diversity of helper T lymphocytes. Nature 383:787-793
5. O'Garra A (1998) Cytokines induce the development of functionally heterogeneous T helper cell subsets. Immunity 8:275-283
6. Hilkens CM, Kalinski P, de Boer M, Kapsenberg ML (1997) Human dendritic cells require exogenous interleukin-12-inducing factors to direct the development of naive T-helper cells toward the Th1 phenotype. Blood 90:1920-1926
7. Macatonia SE, Hsieh CS, Murphy KM, O'Garra A (1993) Dendritic cells and macrophages are required for Th1 development of $CD4^+$ T cells from alpha beta TCR transgenic mice: IL-12 substitution for macrophages to stimulate IFN-gamma production is IFN-gamma-dependent. Int Immunol 5:1119-1128
8. Banchereau J, Bazan F, Blanchard D, Briere F, Galizzi JP, van Kooten C, Liu YJ, Rousset F, Saeland S (1994) The CD40 antigen and its ligand. Annu Rev Immunol 12:881-922
9. Armitage RJ, Fanslow WC, Strockbine L, Sato TA, Clifford KN, Macduff BM, Anderson DM, Gimpel SD, Davis-Smith T, Maliszewski CR et al (1992) Molecular and biological characterization of a murine ligand for CD40. Nature 357:80-82
10. Korthauer U, Graf D, Mages HW, Briere F, Padayachee M, Malcolm S, Ugazio AG, Notarangelo LD, Levinsky RJ, Kroczek RA (1993). Defective expression of T-cell CD40 ligand causes X-linked immunodeficiency with hyper-IgM. Nature 361:539-541
11. Cella M, Scheidegger D, Palmer-Lehmann K, Lane P, Lanzavecchia A, Alber G (1996) Ligation of CD40 on dendritic cells triggers production of high levels of interleukin-12 and enhances T cell stimulatory capacity: T-T help via APC activation. J Exp Med 184:747-752

12. Koch F, Stanzl U, Jennewein P, Janke K, Heufler C, Kampgen E, Romani N, Schuler G (1996) High level IL-12 production by murine dendritic cells: upregulation via MHC class II and CD40 molecules and downregulation by IL-4 and IL-10. J Exp Med 184:741-746
13. Macatonia SE, Hosken NA, Litton M, Vieira P, Hsieh CS, Culpepper JA, Wysocka M, Trinchieri G, Murphy KM, O' Garra A (1995) Dendritic cells produce IL-12 and direct the development of Th1 cells from naive CD4+ T cells. J Immunol 154:5071-5079
14. Sousa CR, Hieny S, Scharton-Kersten T, Jankovic D, Charest H, Germain RN, Sher A (1997) In vivo microbial stimulation induces rapid CD40 ligand-independent production of interleukin 12 by dendritic cells and their redistribution to T cell areas. J Exp Med 86:1819-1829
15. Shortman K, Vremec D, Corcoran LM, Georgopoulos K, Lucas K, Wu L (1998) The linkage between T-cell and dendritic cell development in the mouse thymus. Immunol Rev 165:39-46
16. Inaba K, Inaba M, Deguchi M, Hagi K, Yasumizu R, Ikehara S, Muramatsu S, Steinman RM (1993). Granulocytes, macrophages, and dendritic cells arise from a common major histocompatibility complex class II-negative progenitor in mouse bone marrow. Proc Natl Acad Sci USA 90:3038-3042
17. Schreurs MW, Eggert AA, de Boer AJ, Figdor CG, Adema GJ (1999) Generation and functional characterization of mouse monocyte-derived dendritic cells. Eur J Immunol 29:2835-2841
18. Pulendran B, Lingappa J, Kennedy MK, Smith J, Teepe M, Rudensky A, Maliszewski CR, Maraskovsky E (1997) Developmental pathways of dendritic cells in vivo: distinct function, phenotype, and localization of dendritic cell subsets in FLT3 ligand-treated mice. J Immunol 159:2222-2231
19. Inaba K, Pack M, Inaba M, Sakuta H, Isdell F, Steinman RM (1997) High levels of a major histocompatibility complex II-self peptide complex on dendritic cells from the T cell areas of lymph nodes. J Exp Med 186:665-672
20. Randolph GJ, Inaba K, Robbiani DF, Steinman RM, Muller WA (1999) Differentiation of phagocytic monocytes into lymph node dendritic cells in vivo. Immunity 11:753-761
21. Sallusto F, Lanzavecchia A (1994) Efficient presentation of soluble antigen by cultured human dendritic cells is maintained by granulocyte/macrophage colony-stimulating factor plus interleukin 4 and downregulated by tumor necrosis factor alpha. J Exp Med 179:1109-1118
22. Romani N, Gruner S, Brang D, Kampgen E, Lenz A, Trockenbacher B, Konwalinka G, Fritsch PO, Steinman RM, Schuler G (1994) Proliferating dendritic cell progenitors in human blood. J Exp Med 180:83-93
23. Lennert K, Kaiserling E, Muller-Hermelink HK (1975) T-associated plasma-cells. Lancet 1:1031-1032
24. Facchetti F, de Wolf-Peeters C, Mason DY, Pulford K, van den Oord JJ, Desmet VJ (1988) Plasmacytoid T cells. Immunohistochemical evidence for their monocyte/macrophage origin. Am J Pathol 133:15-21
25. Rissoan MC, Soumelis V, Kadowaki N, Grouard G, Briere F, de Waal Malefyt R,Liu YJ (1999) Reciprocal control of T helper cell and dendritic cell differentiation. Science 283:1183-1186
26. Grouard G, Rissoan MC, Filgueira L, Durand I, Banchereau J, Liu YJ (1997) The enigmatic plasmacytoid T cells develop into dendritic cells with interleukin (IL)-3 and CD40-ligand. J Exp Med 185:1101-1111
27. Olweus J, BitMansour A, Warnke R, Thompson PA, Carballido J, Picker LJ, Lund-Johansen F (1997) Dendritic cell ontogeny: a human dendritic cell lineage of myeloid origin. Proc Natl Acad Sci USA 94:12551-12556
28. O' Doherty U, Peng M, Gezelter S, Swiggard WJ, Betjes M, Bhardwaj N, Steinman RM (1994) Human blood contains two subsets of dendritic cells, one immunologically mature and the other immature. Immunology 82:487-493
29. Strobl H, Scheinecker C, Riedl E, Csmarits B, Bello-Fernandez C, Pickl WF, Majdic O, Knapp W (1998) Identification of CD68+lin- peripheral blood cells with dendritic precursor characteristics. J Immunol 161:740-748

30. Kohrgruber N, Halanek N, Groger M, Winter D, Rappersberger K, Schmitt-Egenolf M, Stingl G, Maurer D (1999) Survival, maturation, and function of CD11c⁻ and CD11c⁺ peripheral blood dendritic cells are differentially regulated by cytokines. J Immunol 163:3250-9325
31. Robinson SP, Patterson S, English N, Davies D, Knight SC, Reid CD (1999) Human peripheral blood contains two distinct lineages of dendritic cells. Eur J Immunol 29:2769-2778
32. Sorg RV, Kogler G, Wernet P (1999) Identification of cord blood dendritic cells as an immature CD11c- population. Blood 93:2302-2307.
33. Res PC, Couwenberg F, Vyth-Dreese FA, Spits H (1999) Expression of pTalpha mRNA in a committed dendritic cell precursor in the human thymus. Blood 94:2647-2657
34. Kadowaki N, Antonenko S, Ho S, Rissoan M-C, Soumelis V, Porcelli SA, Lanier LL, Liu YJ (2000) Human dendritic cell subsets DC1 and DC2 induce NKT cell differentiation into IFN-γ–producing NKT1 versus IL-4–producing NKT2 cells (in preparation)
35. Peguet-Navarro J, Dalbiez-Gauthier C, Rattis FM, Van Kooten C, Banchereau J, Schmitt D (1995) Functional expression of CD40 antigen on human epidermal Langerhans cells. J Immunol 1995 155:4241-4247
36. Caux C, Massacrier C, Vanbervliet B, Dubois B, Van Kooten C, Durand I, Banchereau J (1994) Activation of human dendritic cells through CD40 cross-linking. J Exp Med 180:1263-1272
37. Pulendran B, Smith JL, Caspary G, Brasel K, Pettit D, Maraskovsky E, Maliszewski CR (1999) Distinct dendritic cell subsets differentially regulate the class of immune response in vivo. Proc Natl Acad Sci USA 96:1036-1041
38. Maldonado-Lopez R, De Smedt T, Michel P, Godfroid J, Pajak B, Heirman C, Thielemans K, Leo O, Urbain J, Moser M (1999) CD8α⁺ and CD8α⁻ subclasses of dendritic cells direct the development of distinct T helper cells in vivo. J Exp Med 189:587-592
39. Hilkens CM, Kalinski P, de Boer M, Kapsenberg ML (1997) Human dendritic cells require exogenous interleukin-12-inducing factors to direct the development of naive T-helper cells toward the Th1 phenotype. Blood 90:1920-1926
40. Kalinski P, Hilkens CM, Wierenga EA, Kapsenberg ML (1999) T-cell priming by type-1 and type-2 polarized dendritic cells: the concept of a third signal. Immunol Today 20:561-567
41. Hartmann G, Weiner GJ, Krieg AM (1999) CpG DNA: a potent signal for growth, activation, and maturation of human dendritic cells. Proc Natl Acad Sci USA 96:9305-9310
42. Jakob T, Walker PS, Krieg AM, von Stebut E, Udey MC, Vogel JC (1999) Bacterial DNA and CpG-containing oligodeoxynucleotides activate cutaneous dendritic cells and induce IL-12 production: implications for the augmentation of Th1 responses. Int Arch Allergy Immunol1 18:457-461
43. Sparwasser T, Koch ES, Vabulas RM, Heeg K, Lipford GB, Ellwart JW, Wagner H (1998) Bacterial DNA and immunostimulatory CpG oligonucleotides trigger maturation and activation of murine dendritic cells. Eur J Immunol 28:2045-2054
44. Jakob T, Walker PS, Krieg AM, Udey MC, Vogel JC (1998) Activation of cutaneous dendritic cells by CpG-containing oligodeoxynucleotides: a role for dendritic cells in the augmentation of Th1 responses by immunostimulatory DNA. J Immunol 161:3042-3049
45. Verdijk RM, Mutis T, Esendam B, Kamp J, Melief CJ, Brand A, Goulmy E (1999) Polyriboinosinic polyribocytidylic acid (poly (I:C)) induces stable maturation of functionally active human dendritic cells. J Immunol 163:57-61
46. King C, Davies J, Mueller R, Lee MS, Krahl T, Yeung B, O'Connor E, Sarvetnick N (1998) TGF-beta1 alters APC preference, polarizing islet antigen responses toward a Th2 phenotype. Immunity 8:601-613
47. Akiba H, Miyahira Y, Atsuta M, Takeda K, Nohara C, Futagawa T, Matsuda H, Aoki T, Yagita H, Okumura K (2000) Critical Contribution of OX40 Ligand to T Helper Cell Type 2 Differentiation in Experimental Leishmaniasis J Exp Med 191:375-380
48. Delespesse G, Ohshima Y, Yang LP, Demeure C, Sarfati M (1999) OX40-Mediated cosignal enhances the maturation of naive human CD4⁺ T cells into high IL-4-producing effectors. Int Arch Allergy Immunol 118:384-386

49. Rogge L, D'Ambrosio D, Biffi M, Penna G, Minetti LJ, Presky DH, Adorini L, Sinigaglia F (1998) The role of Stat4 in species-specific regulation of Th cell development by type I IFNs. J Immunol 161:6567-6574
50. McRae BL, Beilfuss BA, Seventer GA (2000) IFN-beta differentially regulates CD40-induced cytokine secretion by human dendritic cells. J Immunol 164:23-28
51. McRae BL, Semnani RT, Hayes MP, van Seventer GA (1998) Type I IFNs inhibit human dendritic cell IL-12 production and Th1 cell development. J Immunol 160:4298-4304
52. Bartholome EJ, Willems F, Crusiaux A, Thielemans K, Schandene L, Goldman M (1999) IFN-beta interferes with the differentiation of dendritic cells from peripheral blood mononuclear cells: selective inhibition of CD40-dependent interleukin-12 secretion. J Interferon Cytokine Res 19:471-478
53. Bartholome EJ, Willems F, Crusiaux A, Thielemans K, Schandene L, Goldman M (1999) Interferon-beta inhibits Th1 responses at the dendritic cell level. Relevance to multiple sclerosis. Acta Neurol Belg 99:44-52
54. Iwasaki A, Kelsall BL (1999) Freshly isolated Peyer's patch, but not spleen, dendritic cells produce interleukin 10 and induce the differentiation of T helper type 2 cells. J Exp Med 190:229-239
55. Stumbles PA, Thomas JA, Pimm CL, Lee PT, Venaille TJ, Proksch S, Holt PG (1998) Resting respiratory tract dendritic cells preferentially stimulate T helper cell type 2 (Th2) esponses and require obligatory cytokine signals for induction of Th1 immunity. J Exp Med 188:2019-2031
56. Khanna A, Morelli AE, Zhong C, Takayama T, Lu L, Thomson AW (2000) Effects of Liver-Derived Dendritic Cell Progenitors on Th1- and Th2-Like Cytokine Responses In Vitro and In Vivo. J Immunol 164:1346-1354
57. Peter HH, Dallugge H, Zawatzky R, Euler S, Leibold W, Kirchner H. (1980) Human peripheral null lymphocytes. II. Producers of type-1 interferon upon stimulation with tumor cells, Herpes simplex virus and Corynebacterium parvum. Eur J Immunol 10:547-555
58. Ronnblom L, Ramstedt U, Alm GV (1983) Properties of human natural interferon-producing cells stimulated by tumor cell lines. Eur J Immunol 13:471-476
59. Abb J, Abb H, Deinhardt F (1983) Phenotype of human alpha-interferon producing leucocytes identified by monoclonal antibodies. Clin Exp Immunol 52:179-184
60. Perussia B, Fanning V, Trinchieri G. (1985) A leukocyte subset bearing HLA-DR antigens is responsible for in vitro alpha interferon production in response to viruses. Nat Immun Cell Growth Regul 4:120-137
61. Fitzgerald-Bocarsly P, Feldman M, Mendelsohn M, Curl S, Lopez C (1988) Human mononuclear cells which produce interferon-alpha during NK(HSV-FS) assays are HLA-DR positive cells distinct from cytolytic natural killer effectors. J Leukoc Biol. 43:323-334
62. Chehimi J, Starr SE, Kawashima H, Miller DS, Trinchieri G, Perussia B, Bandyopadhyay S (1989) Dendritic cells and IFN-alpha-producing cells are two functionally distinct non-B, non-monocytic HLA-DR^+ cell subsets in human peripheral blood. Immunology 68:488-490
63. Sandberg K, Eloranta ML, Johannisson A, Alm GV (1991) Flow cytometric analysis of natural interferon-alpha producing cells. Scand J Immunol 34:565-576
64. Ferbas JJ, Toso JF, Logar AJ, Navratil JS, Rinaldo Jr CR (1994) $CD4^+$ blood dendritic cells are potent producers of IFN-alpha in response to in vitro HIV-1 infection. J Immunol 152:4649-4662
65. Svensson H, Johannisson A, Nikkila T, Alm GV, Cederblad B (1996) The cell surface phenotype of human natural interferon-alpha producing cells as determined by flow cytometry. Scand J Immunol 44:164-172
66. Siegal FP, Kadowaki N, Shodell M, Fitzgerald-Bocarsly PA, Shah K, Ho S, Antonenko S, Liu YJ (1999) The nature of the principal type 1 interferon-producing cells in human blood. Science 284:1835-1837
67. Cella M, Jarrossay D, Facchetti F, Alebardi O, Nakajima H, Lanzavecchia A, Colonna M (1999) Plasmacytoid monocytes migrate to inflamed lymph nodes and produce large amounts oftype I interferon. Nat Med 5:919-923

ILT Receptors at the Interface Between Lymphoid and Myeloid Cells

M. Cella [1], H. Nakajima [1], F. Facchetti [2], T. Hoffmann [3] and M. Colonna [1]
[1] Basel Institute for Immunology, Basel, Switzerland, [2] Istituto di Anatomia Patologica, Universita' di Brescia, Spedali Civili di Brescia, Brescia Italy, [3] Heinrich Heine Universität, Düsseldorf, Germany

ILT/LIR/MIR Receptors

Immunoglobulin (Ig)-like transcripts (ILT) receptors, also known as monocyte/macrophage Ig-like receptors (MIRs) or leukocyte Ig-like receptors (LIRs) [1, 2], constitute a family of receptors which belong to the immunoglobulin superfamily. They are characterized by either 2 or 4 homologous extracellular C-2 type Ig-like domains or by different transmembrane and cytoplasmic domains. One subset of ILT receptors (ILT2, ILT3, ILT4, ILT5 and LIR8) displays long cytoplasmic tails containing immunoreceptor tyrosine-based inhibitory motifs (ITIMs) and delivers inhibitory signals by recruiting SHP-1 protein tyrosine phosphatase [3-5]. A second subset of ILT receptors (ILT1, ILT7, ILT8 and LIR6) is characterized by the presence of a positively charged amino acid residue within the transmembrane region and by a short cytoplasmic tail with no obvious signal transduction motifs. These receptors mediate cell activation by associating with the gamma chain of Fc receptors (FcRγ) [6]. FcRγ recruits protein tyrosine kinases through a cytoplasmic immunoreceptor tyrosine-based activation motif (ITAM).

ILT receptors (ILTs) are genetically, structurally and functionally related to the killer cell Ig-like receptors (KIRs). KIRs are expressed on natural killer (NK) cells and on subsets of T cells, and bind to MHC class I molecules, thus controlling NK cell-mediated cytotoxicity [7, 8]. ILTs, as compared to KIRs, have a different cellular distribution on cells of the hematopoietic system. Some ILT receptors, namely ILT2, are expressed on a subset of NK and T cells, on all peripheral blood B cells and monocytes, on dendritic cells (DCs) and on macrophages [4, 9]. Other ILT receptors, such as ILT3 and ILT4, have a more restricted expression pattern, since they are only found on cells of the myeloid lineage, such as monocytes, DCs and macrophages [1, 3, 4, 9, 10]. Molecular binding experiments performed with ILT-Fc fusion proteins or soluble MHC class I tetramers, have shown that ILT2 and ILT4, like the KIRs, directly interact with HLA class I molecules. This finding is consistent with the idea that not only NK cells [11], but also other cell types, i.e. macrophages, can monitor the level of HLA class I expression, displaying cytolytic activity against "unhealthy" HLA class I-low target cells. While ILT2 and ILT4 bind HLA class I molecules with a broad specificity, the natural ligands of other ILT receptors, including ILT1, ILT3 and ILT5, are still unknown. Class-I like molecules and non-classical class I molecules are likely candidates [1].

Cell Surface Expression of ILT Receptors is Tightly Regulated During Cell Development

The level of ILT receptors of the inhibitory subset is strictly regulated during development of B cells [12] and myeloid cells. In bone marrow, ILT2 is expressed at low levels on CD34⁺ CD10⁺ CD19⁺ pre-B I cells; higher levels are progressively acquired by pre-B II and immature B cells, with the highest expression detectable on CD10⁻ CD19⁺ mature B cells (Fig.1) [13].

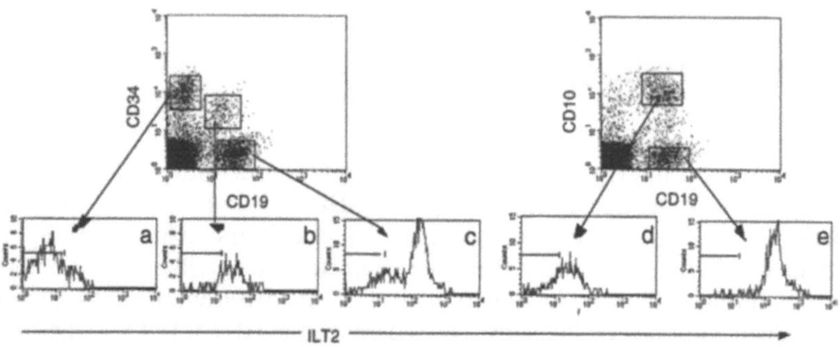

Fig. 1. ILT2 expression on human bone marrow B cells is upregulated during the development of B cells. CD19⁺CD34⁺ cells represent Pre-BI B cell precursors (b), CD19⁺ CD10⁺ cells include Pre-B II and immature B cells (c,d). CD19⁺CD10⁻ cells correspond to mature B cells (e).

A similar behavior is also observed for other inhibitory receptors restricted to cells of the myeloid lineage. The level of expression of ILT5 on monocytes and on granulocytes increases in parallel with that of CD14 and CD16 respectively (Fig. 2). Thus, it appears that a high expression of inhibitory receptors is necessary for the precise tuning of the effector function of mature cells [14].

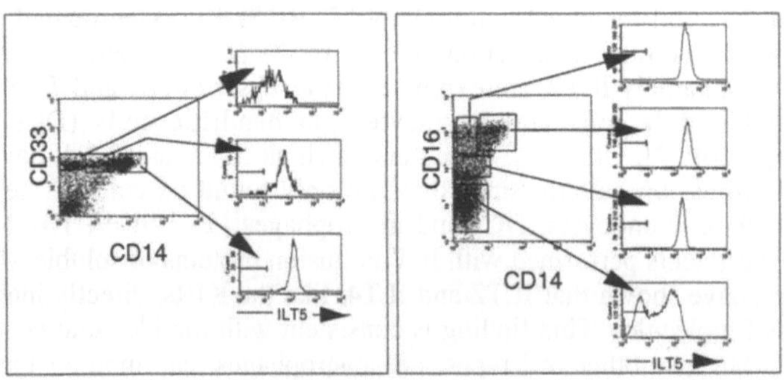

Fig. 2. ILT5 expression is increased on monocytes and granulocytes during maturation. Human bone marrow cells were analyzed by flow cytometry for expression of ILT5 and CD14 and CD33, or CD16 and CD14. Monocytes (left panel) and granulocytes (right panel) were gated on the basis of their size. The most intense ILT5 expression is observed on CD14 ^{high} mature monocytes and on CD16 ^{high} CD14⁻ and CD16 ^{high} CD14 ^{dim} mature granulocytes.

Accordingly, ILT2 and ILT4, when engaged by the appropriate HLA class I molecules, inhibit the secretion of cytokines and chemokines by mature circulating monocytes and DCs. In particular, the production of bioactive IL-12p75 by CD40L-activated DCs is sharply reduced when ILT2 and ILT4 are engaged by HLA class I molecules expressed on the same cells which express CD40L (Fig. 3, left panel). Such inhibition is partially reverted by anti-HLA class I or anti-ILT2 and anti-ILT4 monoclonal antibodies which prevent the interaction between the inhibitory receptors and their ligands (Fig. 3, right panel.).

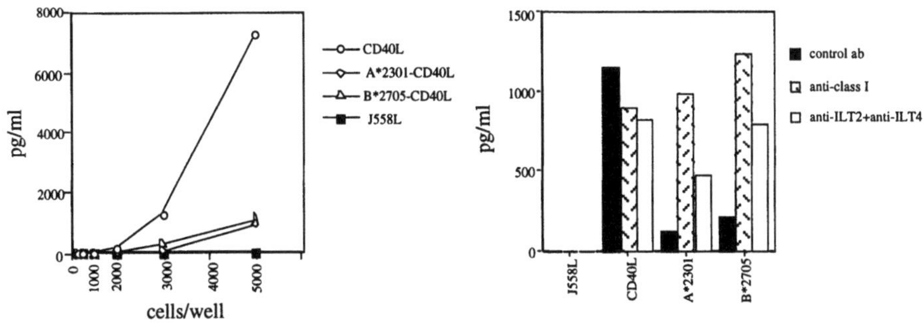

Fig. 3. Engagement of the inhibitory receptors ILT2 and ILT4 on DCs by HLA class I molecules reduces IL-12p75 production triggered by CD40L.
Left panel: graded number of mock-transfected J558L cells, of CD40L-transfected J558L cells, and of A*2301- or B*2705-CD40L-double transfected J558L cells, were cocultured with DCs for 40 hours. Supernatants were collected and the content of IL-12p75 was measured by ELISA. Right panel: anti-HLA class I antibodies, or anti-ILT2 and anti-ILT4 antibodies, can partially revert the inhibition of IL-12 production. The antibodies used in the assay have been shown to block the interaction between class I molecules and the inhibitory receptors. All the transfectants expressed comparable levels of CD40L.

ILTs are Differentially Expressed by Subsets of Primary DCs

ILT3 is an inhibitory receptor that is rapidly internalized upon crosslinking, suggesting that it may interact with a multivalent ligand and may have a role in antigen presentation. ILT3 is expressed only on cells of myeloid origin, such as monocytes, DCs and macrophages derived from monocytes upon *in vitro* culture [15], as well as $CD83^+$ mature DCs in human blood [3]. In addition, ILT3 is also expressed on $CD14^-$, $HLA\text{-}DR^{bright}$ primary DCs. Analysis of this latter population with a monoclonal antibody specific for ILT1 [6], allowed the identification of two subsets of MHC class II positive DCs [16]. The first subset, $ILT3^+/ILT1^+$ DCs, expresses the classical myeloid markers CD33, CD13 and CD11c and possesses a typical veiled morphology when cultured for a few hours in GM-CSF and IL-4. This subset may contain either immature DCs, which migrate from the bone marrow to the tissues where they have to be continuously seeded, or mature DCs, which directly access secondary lymphoid organs from blood to maintain tolerance [17, 18].

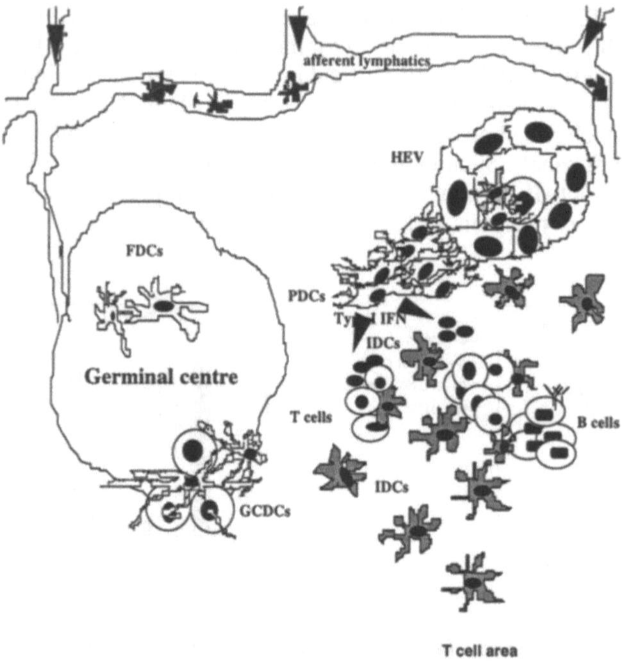

Fig. 4. Plasmacytoid DCs enter into the lymph node through HEVs.
Plasmacytoid DCs (PDCs) enter the lymph node during inflammation attracted by IP10 and Mig chemokines produced by High Endothelial Venules (HEVs). Here they can also meet activated T cells expressing CD40L, undergo a further maturation step and produce high amounts of Type I IFN. Type I IFN can act on surrounding DCs of the T cell areas such as Interdigitating DCs (IDCs), or on antigen activated T cells. Germinal Centre DCs (GCDCs), Follicular DCs (FDCs).

The second subset, ILT3$^+$/ILT1$^-$ DCs, which requires the presence of IL-3 in order to survive in vitro, shows a plasma cell-like morphology, lacks myeloid markers and expresses high levels of CD36, CD68 and IL-3 receptor alpha chain [16, 19-22]. Cells with similar characteristics were previously identified in inflamed lymph nodes, especially around and in the lumen of the high endothelial venules (HEVs) and were termed "plasmacytoid monocytes" by pathologists [23]. In peripheral blood, ILT3$^+$/ILT1$^-$ DCs express high levels of CD62L, which mediates adhesion to endothelium of HEVs. They also exhibit the chemokine receptor CXCR3, which directs their migration to the lymph nodes when HEVs are stimulated by inflammatory cytokines, such as TNF-α and IFN-γ and produce IP10 and Mig chemokines. Thus ILT3$^+$/ILT1$^-$ plasmacytoid DCs possess a unique migratory pathway (Fig. 4), since they access secondary lymphoid organs directly from the blood stream through HEVs, in contrast to DCs mobilized from peripheral tissues, which are believed to enter lymph nodes through afferent lymph [24-26]. ILT3$^+$/ILT1$^-$ plasmacytoid DCs have also the unique functional property of producing high amounts of Type I IFN upon stimulation with influenza virus or CD40L. Thus, they correspond to the previously described natural interferon-producing cells (IPC), which are triggered by many viruses accessing the blood [16, 27, 28]. ILT3$^+$/ILT1$^-$ plasmacytoid DCs, which are likely to be the immature resident DCs of peripheral blood, undergo a strong maturation process when stimulated with CD40L. The production of both bioactive IL-12p75 and Type I interferon, in these circumstances, drives a strong Th1 response, which is remarkably important in the generation of immune responses to viruses. In addition, the presence of Type I IFN, at high concentration in the lymph node, can play a critical role not only in protecting other DCs from the lethal effect of cytophatic viruses, but also in promoting survival of antigen activated T cells, with particular regard to $CD8^+$ cytotoxic T cells [29, 30].

Acknowledgements

The Basel Institute for Immunology was founded and is supported by F. Hoffmann-La Roche Ltd., Basel, Switzerland. The authors thank Dr. Rachel Ettinger and Dr. G. Iezzi for critical reading of the manuscript.

References

1. Colonna M, Navarro F, Lopez-Botet M (1999) A novel family of inhibitory receptors for HLA class I molecules that modulate function of lymphoid and myeloid cells. Curr Top Microbiol Immunol 244:115-122
2. Long EO (1999) Regulation of immune responses through inhibitory receptors. Annu Rev Immunol 17:875-904
3. Cella M, Dohring C, Samaridis J, Brockaus M, Lanzavecchia A, Colonna M (1997) A novel inhibitory receptor (ILT3) expressed on monocytes, macrophages, and dendritic cells involved in antigen processing. J Exp Med 185:1743-1751
4. Colonna M, Navarro F, Bellon T, Llano M, Garcia P, Samaridis J, Angman L, Cella M, Lopez Botet M (1997) A common inhibitory receptor for major histocompatibility complex class I molecules on human lymphoid and myelomonocytic cells [see comments]. J Exp Med 186:1809-1818
5. Cosman D, Fanger N, Borges L, Kubin M, Chin W, Peterson L, Hsu ML (1997) A novel immunoglobulin superfamily receptor for cellular and viral MHC class I molecules. Immunity 7:273-282
6. Nakajima H, Samaridis J, Angman L, Colonna M (1999) Human myeloid cells express an activating ILT receptor (ILT1) that associates with Fc receptor gamma-chain. J Immunol 162:5-8
7. Lanier LL (1998) NK cell receptors. Annu Rev Immunol 16:359-393
8. Lanier LL (1998) Follow the leader: NK cell receptors for classical and nonclassical MHC class I. Cell 92:705-707
9. Navarro F, Llano M, Bellon T, Colonna M, Geraghty DE, Lopez-Botet M (1999) The ILT2(LIR1) and CD94/NKG2A NK cell receptors respectively recognize HLA-G1 and HLA-E molecules co-expressed on target cells. Eur J Immunol 29:277-283
10. Colonna M, Samaridis J, Cella M, Angman L, Allen RL, O'Callaghan CA, Dunbar R, Ogg GS, Cerundolo V, Rolink A (1998) Human myelomonocytic cells express an inhibitory receptor for classical and nonclassical MHC class I molecules. J Immunol 160:3096-3100
11. Ljunggren HG, Karre K (1990) In search of the 'missing self': MHC molecules and NK cell recognition. Immunol Today 11:237-244
12. Ghia P, ten Boekel E, Rolink AG, Melchers F (1998) B-cell development: a comparison between mouse and man. Immunol Today 19:480-485

13. Banham AH, Colonna M, Cella M, Micklem KJ, Puldorf K, Willis AC, Mason DY (1999) Identification of the CD85 antigen as ILT2, an inhibitory MHC class I receptor of the immunoglobulin superfamily. J Leukoc Biol 65:841-845
14. Kubagawa H, Chen CC, Ho LH, Shimada TS, Gartland L, Mashburn C, Uehara T, Ravetch JV, Cooper MD (1999) Biochemical nature and cellular distribution of the paired immunoglobulin-like receptors, PIR-A and PIR-B. J Exp Med 189:309-318
15. Sallusto F, Lanzavecchia A (1994) Efficient presentation of soluble antigen by cultured human dendritic cells is maintained by granulocyte/macrophage colony-stimulating factor plus interleukin 4 and downregulated by tumor necrosis factor alpha. J Exp Med 179:1109-1118
16. Cella M, Jarrossay D, Facchetti F, Alebardi O, Nakajima H, Lanzavecchia A, Colonna M (1999) Plasmacytoid monocytes migrate to inflamed lymph nodes and produce large amounts of type I interferon. Nat Med 5:919-923
17. O'Doherty U, Steinman RM, Peng M, Cameron PU, Gezelter S, Kopeloff I, Swiggard WJ, Pope M, Bhardwaj N (1993) Dendritic cells freshly isolated from human blood express CD4 and mature into typical immunostimulatory dendritic cells after culture in monocyte-conditioned medium. J Exp Med 178:1067-1076
18. O'Doherty U, Peng M, Gezelter S, Swiggard WJ, Betjes M, Bhardwaj N, Steinman RM (1994) Human blood contains two subsets of dendritic cells, one immunologically mature and the other immature. Immunology 82:487-493
19. Strobl H, Scheinecker C, Riedl E, Csmarits B, Bello-Fernandez C, Pickl WF, Majdic O, Knapp W (1998) Identification of CD68+lin- peripheral blood cells with dendritic precursor characteristics. J Immunol 161:740-748
20. Grouard G, Rissoan MC, Filgueira L, Durand I, Banchereau J, Liu YJ (1997) The enigmatic plasmacytoid T cells develop into dendritic cells with interleukin (IL)-3 and CD40-ligand. J Exp Med 185:1101-1111
21. Olweus J, BitMansour A, Warnke R, Thompson PA, Carballido J, Picker LJ, Lund-Johansen F (1997) Dendritic cell ontogeny: a human dendritic cell lineage of myeloid origin. Proc Natl Acad Sci U S A 94:12551-12556
22. Grouard G, Durand I, Filgueira L, Banchereau J, Liu YJ (1996) Dendritic cells capable of stimulating T cells in germinal centres. Nature 384:364-367
23. Facchetti F, de Wolf-Peeters C, van den Oord JJ, Meijer CJ, Pals ST, Desmet VJ (1989) Anti-high endothelial venule monoclonal antibody HECA-452 recognizes plasmacytoid T cells and delineates an "extranodular" compartment in the reactive lymph node. Immunol Lett 20:277-281
24. Dieu MC, Vanbervliet B, Vicari A, Bridon JM, Oldham E, Ait-Yahia S, Briere F, Zlotik A, Lebecque S, Caux C (1998) Selective recruitment of immature and mature dendritic cells by distinct chemokines expressed in different anatomic sites. J Exp Med 188:373-386
25. Sozzani S, Allavena P, D'Amico G, Luini W, Bianchi G, Kataura M, Imai T, Yoshie O, Bonecchi R, Mantovani A (1998) Differential regulation of chemokine receptors during dendritic cell maturation: a model for their trafficking properties. J Immunol 161: 1083-1086
26. Sallusto F, Schaerli P, Loetscher P, Schaniel C, Lenig D, Mackay CR, Qin S, Lanzavecchia A (1998) Rapid and coordinated switch in chemokine receptor expression during dendritic cell maturation. Eur J Immunol 28:2760-2769
27. Perussia B, Fanning V, Trinchieri G (1985) A leukocyte subset bearing HLA-DR antigens is responsible for in vitro alpha interferon production in response to viruses. Nat Immun Cell Growth Regul 4:120-137
28. Siegal FP, Kadowaki N, Shodell M, Fitzgerald-Bocarsly PA, Shah K, Ho S, Antonenko S, Liu YJ (1999) The nature of the principal type 1 interferon-producing cells in human blood. Science 284:1835-1837
29. Tough DF, Borrow P, Sprent J (1996) Induction of bystander T cell proliferation by viruses and type I interferon in vivo. Science 272:1947-1950
30. Marrack P, Kappler J, Mitchell T (1999) Type I Interferons Keep Activated T Cells Alive. J Exp Med 189:521-530

Functional Subsets of Memory T Cells Identified by CCR7 Expression

F. Sallusto*, A. Langenkamp*, J. Geginat§ and A. Lanzavecchia§
*Basel Institute for Immunology, Grenzacherstrasse 487, CH-4005 Basel, Switzerland
§Institute of Research in Biomedicine, Via Vela 6, CH-6500 Bellinzona, Switzerland

Introduction

When naive T cells home to lymph nodes, they first roll along high endothelial venules using CD62L, allowing the chemokine receptor CCR7 to engage its ligand, SLC, displayed by endothelial cells. The CCR7-SLC interaction activates integrins that promote firm adhesion and transmigration of the T cells into the lymph node (Butcher and Picker, 1996; Cyster, 1999).

Effector T cells, which are generated in large numbers at the peak of the primary response, acquire together with effector function a different homing capacity (Mackay, 1993). Effector cells home to inflamed tissues using specific sets of selectin ligands, integrins and chemokine receptors. For instance Th1 express the receptors for E and P selectins PSGL-1/CLA, the receptors for inflammatory chemokines CCR5 and CXCR3, as well as high levels of integrins such as LFA-1 and VLA-4 (Austrup et al., 1997; Sallusto et al., 1998).

After the primary response, a small number of primed T lymphocytes persists for a lifetime. These memory cells can confer protection to a secondary challenge and give rise to a more rapid and effective recall response (Ahmed and Gray, 1996; Zinkernagel et al., 1996). The nature of the cells that harbour immunological memory has been an unresolved issue for a long time. Early studies by Mackay and coworkers demonstrated that unlike naive T cells, memory cells home to peripheral tissues (Mackay et al., 1990). However, some memory T cells must also reach the lymph nodes in order to mount secondary proliferative responses. We have recently identified two subsets of memory T cells that differ for both effector function and migratory capacity and are responsible for the two facets of secondary responses.

Memory T Cell Subsets Within the CD4 Compartment

Human naïve and memory T cells can be distinguished according to their expression of CD45RA and CD45R0 isoforms (Michie et al., 1992). We found that CCR7 is expressed not only on all CD45RA$^+$ naïve T cells, but also on a fraction of CD45RA$^-$ memory cells (Sallusto et al., 1999). When these subsets were sorted and immediately tested for their capacity to produce cytokines in response to a polyclonal stimulation, we found that only CCR7$^-$ memory cells produced effector cytokines such as IFN-γ, IL-4 and IL-5, while CCR7+ memory T cells produced only IL-2 as naïve T cells (Fig. 1).

Proliferative responses to tetanus toxoid (TT) could be measured in both CCR7+ and CCR7- memory subsets, but not in naïve T cells, while IFN-γ was produced in response to TT only by the CCR7- subset (Sallusto et al., 1999). Furthermore, while the magnitude of the proliferative response increased after boosting with TT in both subsets, the relative proportion did not change significantly, indicating that the relative size of these pools is maintained at different times during the immune response by homeostatic mechanisms.

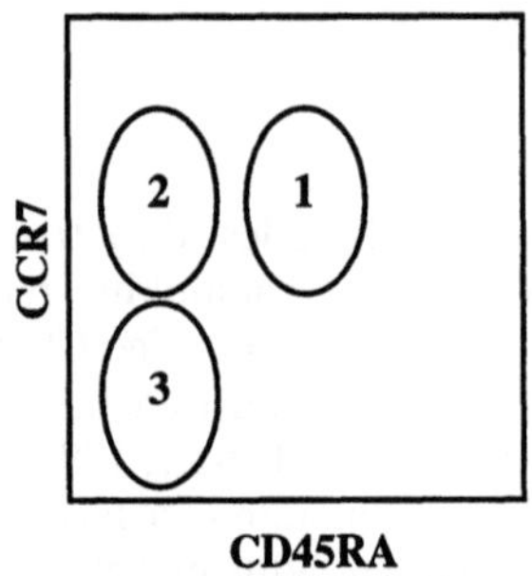

	Subset 1 T_N	Subset 2 T_{CM}	Subset 3 T_{EM}
L-selectin	++	+	-/+
CCR1	-	-	+
CCR3	-	-	+
CCR5	-	-	+
CCR4	-	+	+
CCR6	-	+	+
CXCR3	-	+	+
LFA-1	+	++	+++
VLA-4	-	-	+
IL-2	++	++	+
IFN-γ	-	-	+
IL-4	-	-	+
IL-5	-	-	+
Response to Ag	+	++	+++
DC activation	+	+++	+++
Recall response to TT	-	++	++

Fig. 1. Subsets of CD4 T cells identified according to CD45RA and CCR7 expression.

The CCR7+ and CCR7- subsets also differ for the expression of several surface markers (Fig. 1). As naïve T cells, CCR7+ memory T cells express high levels of L-selectin. However, they have increased levels of LFA-1 and express CCR4, CCR6 and CXCR3 on different proportions of cells. In contrast, CCR7$^-$ T cells express high levels of integrins, as well as receptors for inflammatory chemokines CCR1, CCR3 and CCR5. Because of the differences in homing receptor expression and effector function we named the CCR7$^+$ non-effector cells "central memory" (T_{CM}) and the CCR7$^-$ cells, endowed with Th1 or Th2 effector function, "effector memory" (T_{EM}) T cells (Sallusto et al., 1999).

T_{CM} and T_{EM} Subsets Within the CD8 Compartment

Within the CD8 compartment four subsets of cells could be defined based on CD45RA and CCR7 expression: 1. naïve (CD45RA$^+$, CCR7$^+$), 2. T_{CM} (CD45RA, CCR7$^+$), 3. and 4. two subsets of T_{EM} (CD45RA$^-$, CCR7$^-$ and CD45RA$^+$, CCR7$^-$, respectively) (Sallusto et al., 1999) (Fig. 2). Subset 4 comprises a subset of terminally differentiated cells characterised by the lack of proliferative capacity and CD27 expression as well as a high content of perforin (Hamann et al., 1997).

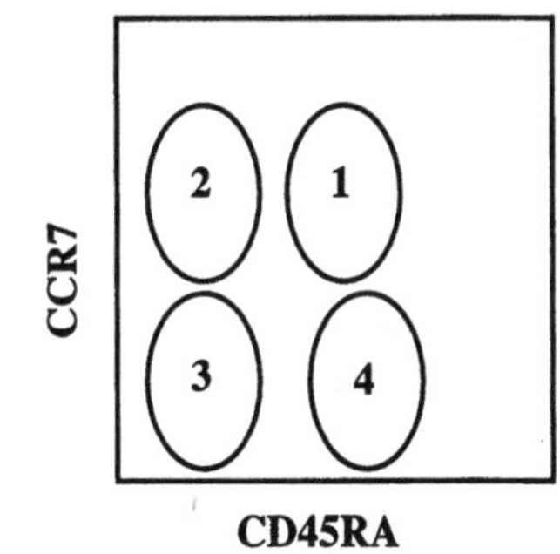

	Subset 1 T_N	Subset 2 T_{CM}	Subset 3 T_{EM}	Subset 4 T_{EM}
L-selectin	++	+	-	-
IL-2	++	+	+	-/+
IFN-γ	-	-	+	+
perforin	-	-	+	++
Tetramer staining	-	+	+	+

Fig. 2. Subsets of CD8 T cells identified according to CD45RA and CCR7 expression.

Antigen-specific memory T cells could be identified by tetramer staining in the three memory subsets. However, their relative proportion differed among individuals, reflecting the time elapsed after antigenic stimulation. When high levels of tetramer positive cells were found, they were mainly of the effector type (subsets 3 and 4). In contrast, when few tetramer positive cells were detected they belonged mainly to subsets 2 and 3. These changes in frequencies of memory cells are consistent with the finding that CD8 responses are particularly dynamic involving large expansion and contraction (Altman et al., 1996). It should be noted in this context that the relative proportion of T_{CM} and T_{EM} cells in peripheral blood do not reflect the size of the two memory pools since T_{CM} but not T_{EM} are present in high numbers in secondary lymphoid organs.

Differentiation of CCR7$^+$ and CCR7$^-$ T Cells from Naïve Cells

Three lines of evidence indicate a linear differentiation process from the naïve stage to a nonpolarized T_{CM} stage and then to a polarised T_{EM} stage (Lanzavecchia and Sallusto, 2000). First, when activated by antigens or mitogens, naïve T cells proliferate and differentiate giving rise to both CCR7$^+$ non-effector cells and CCR7$^-$ effector Th1 or Th2 cells. In these conditions, prolonged stimulation and the presence of polarizing cytokines increase the generation of T_{EM}, while decreasing the yield of T_{CM} (A. Langenkamp et al., manuscript in preparation). Second, when stimulated under the same conditions, T_{CM} proliferate and generate large numbers of CCR7$^-$ effector T cells. Finally, when T_{EM} are stimulated they are capable of undergoing only a limited number of cell divisions, while maintaining the CCR7$^-$ effector phenotype. These experiments indicate that the constitutive expression of CCR7 is maintained as long as effector function is not acquired and is lost as soon as T cells become polarised to Th1 or Th2 or acquire cytotoxic potential.

We suggest that during the primary response antigen specific T cells receive different levels of stimulation that induce their differentiation to either a non-polarized or polarized state. The polarised cells represent the classical effector cells that are found at high levels at the peak of the primary response and die within days or weeks as soon as the antigenic stimulation ceases. It is possible that both subsets are maintained as T_{CM} and T_{EM} for a lifetime by homeostatic mechanisms, for instance the cytokines encountered in the environments of secondary lymphoid organs and peripheral tissues, respectively (Sprent et al., 1997; Tanchot and Rocha, 1998; Unutmaz et al., 1994).

Division of Labour Among Memory T Cells

The finding that immunological memory is harboured within distinct T cell subsets realizes a rationale division of labour. On the one hand, an appropriate frequency of appropriately polarized T_{EM} will be fundamental to confer immediate protection upon a secondary challenge. On the other hand, the frequency of T_{CM} will determine the size of the secondary response. T_{CM} are not only clonally expanded but have also a lower activation threshold as compared to naïve cells, and can potently activate dendritic cells to produce IL-12 (Sallusto et al., 1999). Functionally the T_{CM} subset is likely to be heterogeneous and comprise cells with either positive or negative regulatory functions.

Acknowledgements

The Basel Institute for Immunology was founded and is supported by F. Hoffmann-La Roche Ltd., Basel, Switzerland

References

Ahmed R, Gray D (1996) Immunological memory and protective immunity: understanding their relation. Science 272, 54-60

Altman JD, Moss PAH, Goulder PJR, Barouch DH, McHeyzer-Williams MG, Bell JI, McMichael AJ, Davis MM (1996) Phenotypic analysis of antigen-specific T lymphocytes. Science 274, 94-96

Austrup F, Vestweber D, Borges E, Lohning M, Brauer R, Herz U, Renz H, Hallmann R, Scheffold A, Radbruch A, Hamann A (1997) P- and E-selectin mediate recruitment of T-helper-1 but not T-helper-2 cells into inflamed tissues. Nature 385, 81-83

Butcher EC, Picker LJ (1996) Lymphocyte homing and homeostasis. Science 272, 60-66

Cyster JG (1999) Chemokines and cell migration in secondary lymphoid organs. Science 286, 2098-2102

Hamann D, Baars PA, Rep MH, Hooibrink B, Kerkhof-Garde SR, Klein MR, van Lier RA (1997) Phenotypic and functional separation of memory and effector human $CD8^+$ T cells. J Exp Med 186, 1407-1418

Lanzavecchia A, Sallusto F (2000) From synapses to immunological memory: the role of sustained T cell stimulation. Curr Opin Immunol 12:92-98

Mackay CR (1993) Homing of naive, memory and effector lymphocytes. Curr Opin Immunol 5, 423-427

Mackay CR, Marston WL, Dudler L (1990) Naive and memory T cells show distinct pathways of lymphocyte recirculation. J Exp Med 171, 801-817

Michie CA, McLean A, Alcock C, Beverley PC (1992) Lifespan of human lymphocyte subsets defined by CD45 isoforms. Nature 360, 264-5

Sallusto F, Lanzavecchia A, Mackay CR (1998) Chemokines and chemokine receptors in T-cell priming and Th1/Th2-mediated responses. Immunol Today 19, 568-574

Sallusto F, Lenig D, Forster R, Lipp M, Lanzavecchia A (1999) Two subsets of memory T lymphocytes with distinct homing potentials and effector functions. Nature 401, 708-712

Sprent J, Tough DF, Sun S (1997) Factors controlling the turnover of T memory cells. Immunol Rev 156, 79-85

Tanchot C, Rocha B (1998) The organization of mature T-cell pools. Immunol Today 19, 575-579

Unutmaz D, Pileri P, Abrignani S (1994) Antigen-independent activation of naive and memory resting T cells by a cytokine combination. J Exp Med 180, 1159-1164

Zinkernagel RM, Bachmann MF, Kundig TM, Oehen S, Pirchet H, Hengartner H (1996) On immunological memory. Annu Rev Immunol 14, 333-367 and can potently activate dendritic cells to produce IL-12 (Sallusto et al., 1999). Functionally the T_{CM} subset is likely to be heterogeneous and comprise cells with positive and negative regulatory functions.

Acknowledgement

The Basel Institute for Immunology was founded and is supported by F. Hoffmann-La Roche Ltd, Basel, Switzerland.

References

Ahmed R, Gray D (1996) Immunological memory and protective immunity: understanding their relation. Science 272: 54-60

Altman JD, Moss PAH, Goulder PJR, Barouch DH, McHeyzer-Williams MG, Bell JI, McMichael AJ, Davis MM (1996) Phenotypic analysis of antigen-specific T lymphocytes. Science 274: 94-96

Austrup F, Vestweber D, Borges E, Löhning M, Bräuer R, Herz U, Renz H, Hallmann R, Scheffold A, Radbruch A, Hamann A (1997) P- and E-selectin mediate recruitment of T-helper-1 but not T-helper-2 cells into inflamed tissues. Nature 385: 81-83

Butcher EC, Picker LJ (1996) Lymphocyte homing and homeostasis. Science 272: 60-66

Cyster JG (1999) Chemokines and cell migration in secondary lymphoid organs. Science 286: 2098-2102

Hamann D, Baars PA, Rep MH, Hooibrink B, Kerkhof-Garde SR, Klein MR, van Lier RA (1997) Phenotypic and functional separation of memory and effector human CD8+ T cells. J Exp Med 186: 1407-1418

Lanzavecchia A, Sallusto F (2000) From synapses to immunological memory: the role of sustained T cell stimulation. Curr Opin Immunol 12: 92-98

Mackay CR (1993) Homing of naive, memory and effector lymphocytes. Curr Opin Immunol 5: 423-427

Mackay CR, Marston WL, Dudler L (1990) Naive and memory T cells show distinct pathways of lymphocyte recirculation. J Exp Med 171: 801-817

Michie CA, McLean A, Alcock C, Beverley PC (1992) Lifespan of human lymphocyte subsets defined by CD45 isoforms. Nature 360: 264-5

Sallusto F, Lanzavecchia A, Mackay CR (1998) Chemokines and chemokine receptors in T-cell priming and Th1/Th2-mediated responses. Immunol Today 19: 568-574

Sallusto F, Lenig D, Förster R, Lipp M, Lanzavecchia A (1999) Two subsets of memory T lymphocytes with distinct homing potentials and effector functions. Nature 401: 708-712

Sprent J, Tough DF, Sun S (1997) Factors controlling the turnover of T memory cells. Immunol Rev 156: 79-85

Tanchot C, Rocha B (1998) The organization of mature T-cell pools. Immunol Today 19: 575-579

Unutmaz D, Pileri P, Abrignani S (1994) Antigen-independent activation of naive and memory resting T cells by a cytokine combination. J Exp Med 180: 1159-1164

Zinkernagel RM, Bachmann MF, Kundig TM, Oehen S, Pirchet H, Hengartner H (1996) On immunological memory. Annu Rev Immunol 14: 333-367

Functional Organization of Secondary Lymphoid Organs by the Chemokine System

M. Lipp[1], R. Burgstahler[1], G. Müller[1], V. Pevzner[1] , E. Kremmer[2], E. Wolf[3], and R. Förster[1]

[1] Max-Delbrück-Center for Molecular Medicine, Department of Tumor Genetics and Immunogenetics, Robert-Rössle-Straße 10, 13092 Berlin, Germany

[2] Institute of Molecular Immunology, GSF Research Center for Environment and Health, 81377 Munich, Germany

[3] Institute for Molecular Animal Breeding, Genecenter, Ludwig-Maximilians-University, 81377 Munich, Germany

Introduction

The development of an adaptive immune response requires well coordinated mechanisms in order to navigate circulating immune cells through peripheral tissues and into secondary lymphoid organs. There is strong experimental evidence that chemokines and their receptors are responsible for recruiting cells involved in inflammatory processes as well as for homeostatic control of leukocyte traffic and functional compartmentalization of lymphoid organs. Functional interactions of chemokines with their receptors, which belong to the family of seven-transmembrane-domain proteins signaling through heterotrimeric G proteins, have been shown to be primarily implicated in pathophysiological inflammatory processes (reviewed in Rollins, 1997). Leukocytes are recruited from the blood by locally produced chemokines towards sites of inflammation arising from infections, injury, allergic reactions, arthritis and arteriosclerosis. Several lines of recent evidence suggest that members of the chemokine receptor family are also involved in lymphocyte migration to distinct lymphoid organs and control lymphocyte homeostasis (reviewed in Baggiolini, 1998). Mature B and T lymphocytes continuously recirculate between blood and lymphatics to mediate immune surveillance. In this process they have to interact and to pass specialized endothelia: postcapillary venules in non-lymphoid organs and high endothelial venules (HEVs) in lymph nodes. The recent finding that the chemokine receptor BLR1/CXCR5 is needed for B cell migration into lymphoid follicles (Förster et al., 1996), supported the view that chemokine receptors play an essential role as regulators of adhesion molecules, which, as a whole, allow lymphocyte subsets to navigate to different anatomical locations. These data provided the first experimental evidence that the chemokine system is involved in the functional compartmentalization of lymphoid organs (reviewed in Goodnow and Cyster, 1997). Based on these results chemokines and their receptors can be broadly divided into two functionally distinct categories. On the one hand inflammatory chemokines, induced or up-regulated by inflammatory stimuli, are responsible for recruiting cells involved in acute inflammatory reactions. On the other hand constitutive chemokines, produced in bone marrow, thymus and secondary lymphoid organs, are responsible for the homeostatic control of leukocyte traffic and for driving the encounter of cells that need to interact to generate an immune response (reviewed in Sallusto et al., 1998a).

We and others identified another member of the chemokine receptor family, which is now known as CCR7 (Birkenbach et al., 1993; Burgstahler et al., 1995). Expression of CCR7 has been identified on B cells, T cells, and on activated mature dendritic cells (DC) (Burgstahler et al., 1995; Dieu et al., 1998; Sallusto et al., 1998b; Schweickart et al., 1994). Surface expression of CCR7 is gradually increased on maturing DC and showed a striking resistance to ligand-induced down-regulation (Sallusto et al., 1999b). To further characterize expression and regulation of this chemokine receptor we developed monoclonal antibodies (mAb) specific for this molecule. In this manuscript, we report expression of CCR7 on B cells, T cells and monocytes, but not on neutrophils. Interestingly CCR7 is highly expressed on naive T cells whereas it is only found on a small subset of memory T cells. This expression pattern suggests that CCR7 might be an important homing receptor for naive T cells to secondary lymphoid organs. Indeed, recent data derived from CCR7-deficient, gene-targeted mice support this hypothesis.

Results and Discussion

Generation of Monoclonal Antibodies (mAb) Specific for Human CCR7

In order to obtain mAb specific for extracellular domains of human CCR7 we cloned a GST fusion protein that contains parts of the N-terminus as well as parts of the second extracellular loop of CCR7 (GST-CCR7-NT-ECII). Immunizing rats with this fusion protein we obtained three hybridoma clones (8E8, 3D12, and 5D9) producing antibodies specific for human CCR7 as seen in various differential screening systems. MAb 3D12 binds to an epitope mapping to the N-terminus of human CCR7 and does neither recognize the N-terminus of murine CCR7 expressed as GST fusion protein nor murine CCR7 on transiently transfected HEK293 cells (data not shown). The specificity of mAb 3D12 could be further confirmed in transwell chemotaxis assays placing PBL in the upper and ELC (500ng/ml), which is a ligand for CCR7, in the lower chamber. As can be seen in Fig. 1 pre-incubation of PBL with anti-CCR7 resulted in complete inhibition of ELC-induced chemotaxis.

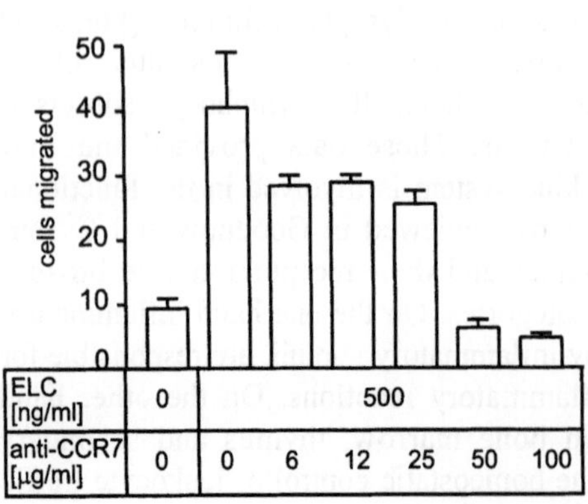

Fig. 1: MAb 3D12 inhibits ELC-induced chemotaxis. PBL were incubated with anti-CCR7 mAb (clone 3D12) at various concentrations and used in transwell chemotaxis assays to assess their chemotactic properties towards ELC. At concentrations >50µg/ml anti-CCR7 mAb completely inhibits ELC-mediated chemotaxis.

CCR7 Expression on Peripheral Blood Leukocytes

To gain insight into the expression pattern of CCR7, we analyzed peripheral blood from four donors by three color flow cytometry. $CD19^+$ B cells were found to be predominantly CCR7-positive (Fig. 2A). In contrast, $CD3^+$ T cells did not express CCR7 uniformly. On average only 57% of peripheral $CD3^+$ cells were found to express CCR7. Monocytes, but not neutrophils, of all four donors expressed CCR7 (Fig. 2A). When analyzing T cells in more detail, we found that the vast majority of naive ($CD45R0^-$) $CD4^+$ and naive $CD8^+$ cells expressed high levels of CCR7 (Fig. 2B). In contrast, a considerably lower proportion of memory T cells expressed CCR7 and expression levels were usually lower than on naive cells (Fig. 2B). As naive T cells enter lymphoid organs via HEV this data explains why T cell areas of all secondary lymphoid organs of CCR7-deficient mice are devoid of T cells (see below).

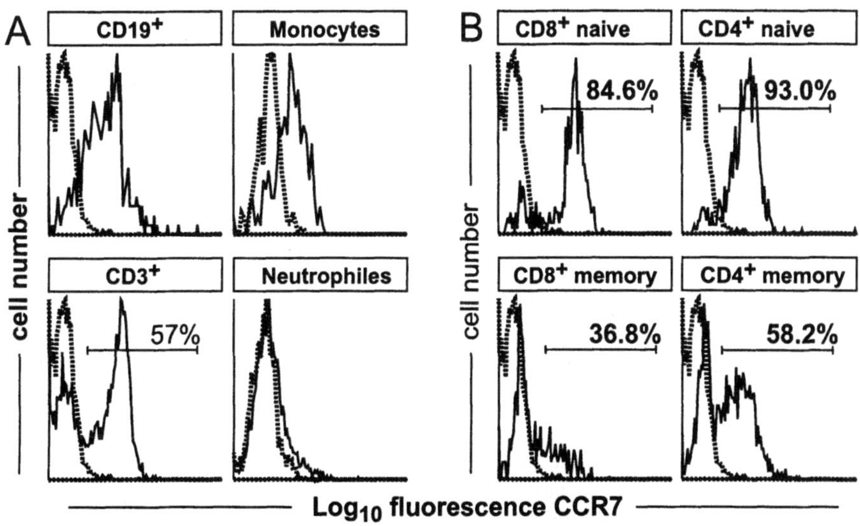

Fig. 2. Expression of CCR7 on peripheral blood leukocytes. (A) Lymphocytes were isolated and expression of CCR7 was analyzed by two color flow cytometry. B cells were identified by anti-CD19 and T cells by anti-CD3 in the FITC channel. Monocytes and neutrophils were analyzed for CCR7 expression by setting appropriate gates on forward and sideward scatter intensities. (B) To determine CCR7 expression on T cell subsets lymphocytes were stained with anti-CD4-FITC (or anti-CD8-FITC), anti-CD45R0-PE and biotinylated anti-CCR7/streptavidin-Cychrome.

Up-regulation of CCR7 on Memory and Naive T Cells Following Stimulation

Upon activation of primary T cells by CD3 crosslinking CCR7 expression was transiently elevated within the first three days of culture. Interestingly, up-regulation of surface CCR7 was more pronounced in the subset of memory T cells (Fig. 3) suggesting that a proportion of memory T cells re-acquires the ability to re-circulate through secondary lymphoid organs. With respect to the function of CCR7 as a homing receptor for memory T cells to lymphoid organs, Sallusto et al.

(1999a) recently demonstrated that CCR7$^+$ memory T cells express high levels of L-selectin (CD62L), which is also required for lymphocytes to pass HEV.

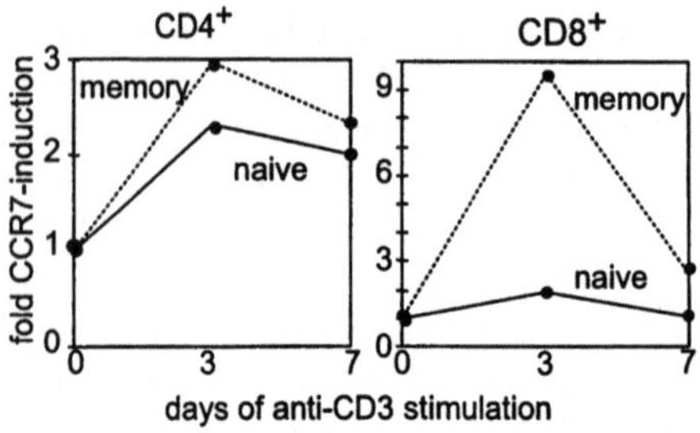

Fig. 3: Up-regulation of CCR7 expression on naive and memory T cells upon anti-CD3 stimulation. T cells were activated with plastic-bound anti-CD3 mAb. At the time points indicated the expression of CCR7 of T cell subpopulations has been determined.

Interestingly, these cells lack immediate effector function, but efficiently stimulated DC and differentiated to effector cells upon TCR stimulation. In contrast, CCR7$^-$ memory T cells express receptors for migration to inflamed tissues like CLA, and displayed immediate effector functions. These results imply that the memory response is mediated by two distinct T cell subsets: tissue seeking effector memory T cells (T_{EM}) can provide immediate protection in inflamed tissue while lymph node seeking central memory T cells (T_{CM}) can provide help for DC and B cells and generate a new wave of effector cells (Sallusto et al., 1999a).

Impaired T Cell Homing in CCR7-deficient Mice

Generating CCR7-deficient mice by gene-targeting we could recently confirm the importance of this chemokine receptor in the process of T cell migration to lymphoid organs. These animals show a severe redistribution of T cells within the body. CCR7$^{-/-}$ mice have highly elevated numbers of naive T cells in the blood, the red pulp of the spleen and the bone marrow, whereas the number of naive cells is severely reduced in all lymph nodes, Peyer's patches (PP) and the splenic white pulp (Förster et al., 1999). In addition, the majority of T cells present in CCR7-mutant mice are displaced in lymphoid organs. In PP T cells do not localize to the inter follicular region (as seen in wild type mice, Fig. 4a) but can be found in the subepithelial dome (Fig. 4d). In lymph nodes isolated from wild types, T cells are almost exclusively found in the medulla (Fig. 4b), which is less densely packed in CCR7-mutant mice (Fig. 4e). In addition, a considerable number of T cells can be identified in the marginal sinus of these mice (Fig. 4e), suggesting that re-circulation of memory T cells via afferent lymphatics is also controlled by CCR7. In the spleen, which is usually 2-3 fold enlarged in mutant mice, a rather complex situation regarding the distribution of T cells can be envisaged. In this organ, the majority of T cells are not found in the periarteriolar lymphoid sheath (PALS) but segregate to the red pulp (Fig. 4f; compare Fig. 4c).

Although it seems clear now that chemokine receptors essentially contribute to lymphoid organ development by directing the migration of lymphoid cells to and through these organs, there is also evidence that CXCR5 is required for establishing the formation of inguinal lymph nodes and PP (Förster et al., 1996).

We recently succeeded to generate a second CXCR5-deficient mouse strain on a different mixed 129-Sv x BALB/c background, in which the *blr*1 gene was replaced by the CRE recombinase using a knock-in strategy (Pevzner et al., unpublished). Most interestingly, these mice exhibited the same developmental deficiencies in lymphoid organogenesis as observed previously. In both strains CXCR5-deficiency interfered with the formation of certain lymph nodes and severely affected the development of Peyer's patches suggesting a direct role of CXCR5 signaling in lymphoid organogenesis. It is known that during fetal lymph node organogenesis in mice, lymph node postcapillary high endothelial venules briefly express the Peyer's patch addressin MAdCAM-1. This allows initial seeding by an unusual lymphocyte population of $CD4^{+}CD3^{-}$ oligolineage progenitors selectively expressing the Peyer's patch homing receptor integrin $\alpha4\beta7$. Recently, it was found that the $CD4^{+}CD3^{-}$ cells are lineage-restricted progenitors that express surface lymphotoxin-β (LTβ) and the chemokine receptor BLR1/CXCR5 (Mebius et al., 1997). They can differentiate into natural killer cells, dendritic antigen-presenting cells, and follicular cells of unknown outcome, but these cells do not become T or B lymphocytes. In view of the necessity of lymphotoxin in lymphoid organ development, it is thought that the novel subset of $CD4^{+}CD3^{-}$ $LT\beta^{+}$ fetal cells is instrumental in the development of lymphoid tissue architecture. Re-evaluation of *blr1* expression during mouse embryogenesis revealed that BLR1-specific mRNA can be detected from day 17.5 p.i. in Northern analysis, but already around day 7.5 p.i. applying PCR-based techniques (Pevzner et al., unpublished).

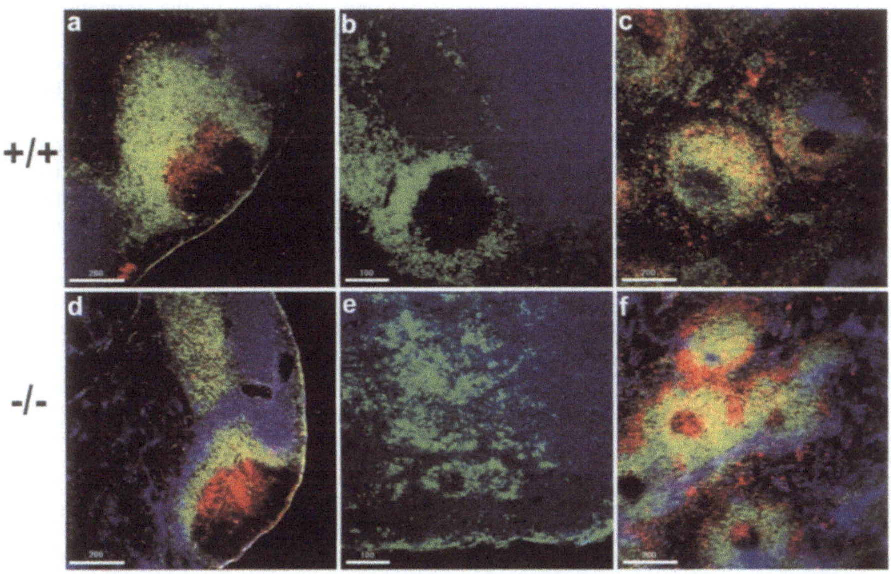

Fig. 4: Misplaced T cells in secondary lymphoid organs of CCR7-deficient mice. Cryosections of Peyer's patches (a + d), mesenteric lymph node (b + e) and spleen (c + f) of wild type (a-c) or CCR7-deficient (d-f) mice were stained with anti-IgD mAb (green) and anti-Thy-1 mAb (blue). In addition sections a, c, d, and f were also stained with anti-IgM mAb (red).

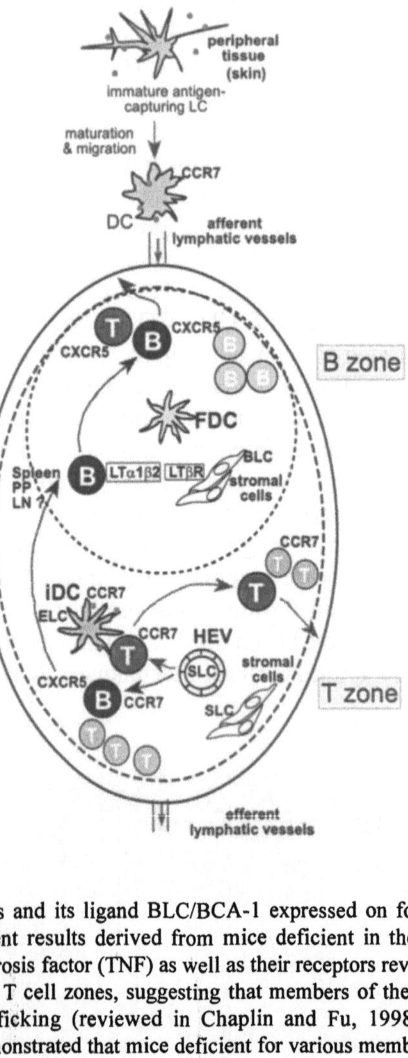

Fig. 5. Lessons from chemokine receptor knock outs: Model of chemokine-directed trafficking of lymphocytes and dendritic cells to and through secondary lymphoid organs during the immune response. (Förster et al., 1996; Förster et al., 1999). B cells and naive T cells enter lymph nodes via high endothelial venules, mediated by binding of the HEV-expressed chemokine SLC to CCR7 on lymphocytes. Once inside the organ, T cells are attracted by the chemokine ELC, which is expressed on interdigitating dendritic cells (iDC) to screen these cells for presented peptides fitting to their antigen receptor. In most cases no specific peptide is presented and the vast majority of T cells leave the lymph node within a few hours. However, if fitting antigen has been identified, T cells start to proliferate. Having entered the T cell-rich area B cells continue to migrate to the B cell follicle. If they are not stimulated by T cells specific for the same antigen, the majority of B cells leaves the follicle within one day. However, if B cells have been activated, they start to proliferate and give rise to the formation of germinal centers within the follicle. The migration of B cells from the T cell rich area of spleen and PP to the follicles is mediated by interaction of CXCR5 on B cells and its ligand BLC/BCA-1 expressed on follicular stroma cells. Remarkably, earlier and recent results derived from mice deficient in the cytokines lymphotoxin (LT) α/β and tumor necrosis factor (TNF) as well as their receptors revealed an impaired architecture of B cell follicles and T cell zones, suggesting that members of the TNF family might be involved in lymphocyte trafficking (reviewed in Chaplin and Fu, 1998). Interestingly, Ngo et al., (1999) recently demonstrated that mice deficient for various members of the TNF/LT family also have an aberrant expression of several chemokines. Reduced levels of chemokines might help to explain how the TNF/LT deficiency affects lymphocyte migration. Once challenged by antigen dendritic cells of the skin get activated and change their expression pattern of chemokine receptors. Several of these receptors get downregulated, whereas CCR7 is increasingly upregulated to allow entry of secondary lymphoid organs via afferent lymphatics. These cells populate the T cell area as iDC.

Conclusion

There is strong experimental evidence that chemokines and their receptors are responsible for recruiting cells involved in inflammatory processes as well as for homeostatic control of leukocyte traffic and functional compartmentalization of lymphoid organs. By analyzing mice deficient for chemokine receptors CXCR5 or CCR7, we could identify both receptors as the principal regulators of lymphoid tissue-specific migration of B and T lymphocytes and dendritic cells. In addition CXCR5 essentially contribute to the development of PP and lymph nodes.

References

Baggiolini M (1998). Chemokines and leukocyte traffic. Nature 392, 565-568.

Birkenbach M, Josefsen K, Yalamanchili R, Lenoir G, and Kieff E (1993). Epstein-Barr virus-induced genes: First lymphocyte-specific G protein-coupled peptide receptors. J. Virol. 67, 2209-2220.

Burgstahler R, Kempkes B, Steube K, and Lipp M (1995). Expression of the chemokine receptor BLR2/EBI-1 is specifically transactivated by Epstein Barr virus nuclear antigen 2. Biochem. Biophys. Res. Com. 215, 737-743.

Chaplin DD, and Fu Y (1998). Cytokine regulation of secondary lymphoid organ development. Curr. Opin. Immunol. 10, 289-297.

Dieu MC, Vanbervliet B, Vicari A, Bridon JM, Oldham E, Ait-Yahia S, Briere F, Zlotnik A, Lebecque S, and Caux C (1998). Selective recruitment of immature and mature dendritic cells by distinct chemokines expressed in different anatomic sites. J. Exp. Med. 188, 373-386.

Förster R, Mattis EA, Kremmer E, Wolf E, Brem G, and Lipp M (1996). A putative chemokine receptor, BLR1, directs B cell migration to defined lymphoid organs and specific anatomic compartments of the spleen. Cell 87, 1037-1047.

Förster R, Schubel A, Breitfeld D, Kremmer E, Renner-Müller I, Wolf E, and Lipp M (1999). CCR7 coordinates the primary immune response by establishing functional microenvironments in secondary lymphoid organs. Cell 99, 23-33.

Goodnow CC, and Cyster JG (1997). Lymphocyte homing: the scent of a follicle. Curr. Biol. 7, 219-222.

Mebius RE, Rennert P, and Weissman IL (1997). Developing lymph nodes collect CD4+CD3-LTbeta+ cells that can differentiate to APC, NK cells, and follicular cells but not T or B cells. Immunity 7, 493-504.

Ngo VN, Korner H, Gunn MD, Schmidt KN, Riminton DS, Cooper MD, Browing JL, Sedgwick JD, and Cyster JG (1999). Lymphotoxin alpha/beta and tumor necrosis factor are required for stromal cell expression of homing chemokines in B and T cell areas of the spleen. J. Exp. Med. 189, 403-412.

Rollins BJ (1997). Chemokines. Blood 90, 909-928.

Sallusto F, Lanzavecchia A, and Mackay CR (1998a). Chemokines and chemokine receptors in T-cell priming and Th1/Th2-mediated responses. Immunol. Today 19, 568-574.

Sallusto F, Schaerli P, Loetscher P, Schaniel C, Lenig D, Mackay CR, Qin S, and Lanzavecchia A (1998b). Rapid and coordinated switch in chemokine receptor expression during dendritic cell maturation. Eur. J. Immunol. 28, 2760-2769.

Sallusto F, Lenig D, Forster R, Lipp M, and Lanzavecchia A (1999a). Two subsets of memory T lymphocytes with distinct homing potentials and effector functions. Nature 401, 708-712.

Sallusto F, Palermo B, Lenig D, Miettinen M, Matikainen S, Julkunen I, Forster R, Burgstahler R, Lipp M, and Lanzavecchia A (1999b). Distinct patterns and kinetics of chemokine production regulate dendritic cell function. Eur. J. Immunol. 29, 1617-1625.

Schweickart VL, Raport CJ, Godiska R, Byers MG, Eddy RJ, Shows TB, and Gray PW (1994). Cloning of human and mouse EBI1, a lymphoid-specific G-protein-coupled receptor encoded on human chromosome 17q12-q21.2. Genomics 23, 643-650.

The Cluster of ABCD Chemokines which Organizes T Cell-dependent B Cell Responses

C. Schaniel, F. Melchers, and A. G. Rolink
Basel Institute for Immunology, Grenzacherstrasse 487, CH-4005 Basel, Switzerland

Introduction

The ordered migration and cooperation of many different lineages of hematopoietic cells is a key requirement for the development and the responses of cells of the immune system and during an immune response. The three major hematopoietic cell types, B lymphocytes, T lymphocytes and antigen presenting cells (APC), originate from pluripotent hematopoietic cells, and develop in the primary lymphoid organs (bone marrow and thymus). Sites for positive immune responses are the secondary lymphoid organs: spleen, lymph nodes (LN), Peyer's patches (PP) and tonsils.

The trafficking of cells of the immune system to discrete compartments of the body and the congregating of these single cells into lymphoid structures in organs are not random processes. Cell migration is controlled by multistep processes which include adhesion of cells, transmigration through cell barriers, chemoattraction and, finally, cell-cell interactions in development and responses (Butcher et al., 1999; Springer, 1994). It has become evident that chemokines and their corresponding seven-transmembrane spanning receptors (for a recent review see Zlotnik et al., 1999) play important roles in the microanatomical homing of lymphoid cells which results in ordered associations of APC, T cells and B cells and, ultimately, in the formation of germinal centers (Melchers et al., 1999).

Here we summarize what is known about a cluster of three ABCD chemokines in mice and man called ABCD (activated B cells and cells of the dendritic type-myeloid cells)-1 [also known as macrophage-derived chemokine (MDC) (Godiska et al., 1997), or stimulated T cell chemotactic protein (STCP-1) (Chang et al., 1997) or DC/B-CK (Ross et al., 1999)], ABCD-2 [thymus and activation-regulated chemokine (TARC) (Imai et al., 1996; Lieberam and Förster, 1999)], and ABCD-3 [fractalkine or neurotactin (Bazan et al., 1997; Pan et al., 1997; Rossi et al., 1998)] which appear to be instrumental in the ordered migration and association of T, B and APC during an immune response to T cell-dependent antigens. We will briefly describe the isolation and structural organization of these chemokines, their expression pattern, and the cell population(s) chemoattracted by them. We will discuss their possible involvement in the organization of T cell-dependent B cell responses and germinal center formation.

Identification and Characterization of the ABCD Chemokines

Three chemokines produced by activated B cells and cells of the dendritic type-myeloid cells have been identified. They are ABCD-1, ABCD-2 and ABCD-3. All three genes are strongly upregulated in mature Sμ-Sε-switched cells generated *in vitro* from pro/pre-B I cells stimulated with anti-CD40 in the presence of interleukin (IL)-4 (Schaniel et al., 1998; 1999). Initially, this allowed the isolation of the gene encoding ABCD-1 from a library of genes expressed differentially in the mature Sμ-Sε-switched cells (Schaniel et al., 1998). Subsequently a second chemokine, ABCD-2, with strong sequence homology to ABCD-1, was found (Schaniel et al., 1999). Both ABCD-1 and -2 belong to the CC chemokine family. ABCD-3, on the other hand, is the only member of the CX_3C chemokine family known today. It contains, besides the chemokine domain, a mucin-like stalk and a transmembrane region (Bazan et al., 1997; Pan et al., 1997; Rossi et al., 1998). The three murine ABCD chemokine genes are clustered on a small genomic fragment within 120 kb, most likely on mouse chromosome 8, since their human homologues have been localized on the synthenic position on human chromosome 16 (Nomiyama et al., 1997; Schaniel et al., 1999). Each ABCD gene is encoded by three exons. The entire ABCD-1 gene is split by a ~1.2 kb and a ~2.7 kb intron (in humans ~1.5 kb and ~2.9 kb), the ABCD-2 gene by a ~0.6 kb and a ~0.5 kb intron (~1.0 kb and ~0.6 kb), and the ABCD-3 gene by a ~5.9 kb and a ~1.5 kb intron (~7.0 kb and ~2.3 kb) (Schaniel et al., 1998; 1999; and unpublished data). In addition, the sequences at the exon-intron boundaries of each mouse and human gene show striking identity.

The ABCD Chemokines are Produced by Dendritic Cells and Activated B Lymphocytes

The expression pattern of ABCD mRNAs was studied by Northern hybridization and/or by semi-quantitative RT-PCR. In all experiments, expression of the individual ABCD chemokine mRNA was compared with the other ABCD chemokine mRNAs, and with the ubiquitously expressed β-actin mRNA.

No message for ABCD-1 and -2 was detectable in brain, liver, kidney, and bone marrow. Relatively little expression of ABCD-1 mRNA was detected in lung, spleen, thymus, and LN. In contrast, low levels of ABCD-2 were only detectable in thymus and LN, hardly in lung, and not at all in spleen (Lieberam and Forster, 1999; Schaniel et al., 1999). ABCD-1, as well as ABCD-2 mRNA, was absent from T cell lines, activated LN T cells, IL-2 activated natural killer (NK) cells and from all B lineage stages of the bone marrow as well as immature and mature B cells of the spleen (Schaniel et al., 1998; 1999). It has been reported for both mice and man, that ABCD-1 message is detectable in T helper cells of type 1 and in even larger amounts in type 2 helper T cells after several rounds of *in vitro* polarization (Galli et al., 2000; Lloyd et al., 2000). In contrast to the human system (Chang et al., 1997; Chantry et al., 1998; Godiska et al., 1997; Rodenburg et al., 1998), we and others have not yet found any detectable message for either ABCD-1 or -2 in *in vitro* grown and activated macrophages (Ross et al., 1999; Schaniel et al., 1998; 1999). However, ABCD-1 protein was detected in alveolar macrophages during an ovalbumin-induced lung allergic inflammation (Gonzalo et al., 1999).

Resting B cells did not produce ABCD-1, ABCD-2, or ABCD-3. High levels of ABCD-1 and -2 message were seen, as expected, in pro/pre-B cell lines stimulated with anti-CD40 and IL-4. ABCD-1 and -2 mRNA were inducible by stimulation with anti-CD40 and IL-4, or by anti-CD40 alone, in mature splenic B cells. T cell-independent stimulation of mature splenic B cells by LPS or by anti-IgM also induced ABCD-1 and -2 mRNA (Table 1; Schaniel et al., 1998; 1999).

Another source for ABCD-1 and -2 production are maturing Langerhans cells and dendritic cells (DC) (Table 1; Schaniel et al., 1998; 1999; Tang and Cyster, 1999; Kanazawa et al., 1999: Lieberam and Förster, 1999). DC (>85% CD11c$^+$) derived from RAG-2 deficient bone marrow by *in vitro* culture in the presence of GM-CSF expressed high levels of ABCD-1 and -2 mRNA, comparable to those seen in anti-CD40 activated splenic B cells. Message for ABCD-1 and -2 was also detectable in freshly isolated DC of several organs. ABCD-1 mRNA was detected in DC from spleen, epidermis, PP and LN, whereas mRNA for ABCD-2 was seen in DC from thymus, LN, PP and lung.

High amounts of ABCD-3 mRNA were readily detectable in brain, kidney and lung, and at lower levels in liver (Bazan et al., 1997; Pan et al., 1997; Rossi et al., 1998; Schaniel et al., 1999). Neurons were identified as the constitutive source of ABCD-3 in the brain (Harrison et al., 1998). ABCD-3 could also be induced in astrocytes by TNF-α and IL-1β stimulation (Maciejewski-Lenoir et al., 1999).

We have found the presence of high levels of ABCD-3 in pro/pre B cells stimulated with anti-CD40 and IL-4 (Schaniel et al., 1999). Low amounts of ABCD-3 message were also found in anti-CD40 activated B cells from the spleen. In contrast to ABCD-1 and -2 production, neither anti-IgM nor LPS induced ABCD-3 production in B cells (Table 1; Schaniel et al., 1999).

DC produce ABCD-3, as they do ABCD-1 and -2 (Table 1). The DC can be either from bone marrow, epidermis, spleen, LN or PP (Kanazawa et al., 1999; Schaniel et al., 1999).

Table 1. Expression pattern of ABCD chemokines

	Resting B	Activated B cells: LPS	anti-IgM	anti-CD40 (+ IL-4)	DC
ABCD-1	-	+	+	+	+
ABCD-2	-	+	+	+	+
ABCD-3	-	-	-	+	+

The Cell Targets for ABCD-1, -2 and -3

We have used an *in vitro* microchamber migration assay to study the chemotactic properties of purified recombinant ABCD-1 and -2, produced in insect cells, as well as of the chemokine portion of ABCD-3 on different cell populations (Schaniel et al., 1998; 1999). No significant migratory response to either ABCD-1 or -2 was observed with bone marrow cells, total thymocytes, primary spleen cells, LN cells (consisting in majority of single CD4$^+$ and CD8$^+$ T lymphocytes) (Schaniel et al., 1998; 1999). Moreover, neither resting nor LPS-activated splenic CD19$^+$ B cells showed any significant migration (Ross et al., 1999; Schaniel et al.,

1998; 1999). IL-2 activated NK cells from mice were also not chemoattracted by ABCD-1 or -2 (Schaniel et al., 1998; 1999). This is in contrast to human ABCD-1 which was shown to attract human IL-2 induced NK cells (Godiska et al., 1997).

Human ABCD-1 was unable to induce migration of neutrophils and eosinophils (Chang et al., 1997; Godiska et al., 1997). The migration of monocytes towards human ABCD-1 remains controversial, as chemoattraction was reported in one (Godiska et al., 1997) but not the other study (Chang et al., 1997).

Differences in migratory capacities were seen between human and mouse myeloid cells. Human ABCD-1 could attract human monocyte-derived DC (Godiska et al., 1997) whereas mouse ABCD-1 was unable to induce migration of bone marrow-derived DC, freshly isolated or short-term *in vitro* cultivated Langerhans cells (Ross et al., 1999).

Both ABCD-1 and -2 are expressed in the thymus. It was suggested that ABCD-1 plays a role in controlling the transitional state from $CD4^+CD8^+$ to $CD4^+$ or $CD8^+$ thymocytes and the further migration out of the thymus, since it attracts the thymocytes of the thymic medulla but not the immature thymocytes of the thymic cortex (Campbell et al., 1999). However, this ABCD-1 responsiveness is apparently not an absolute necessity, since ABCD-1 deficient mice show no disturbance in thymocyte development and migration to the periphery (T. Shimizu, C.S., A.G.R., and F.M., unpublished data). Secondary lymphoid-tissue chemokine (SLC) and Epstein-Barr virus-induced molecule 1 ligand chemokine (ELC) and thymus-expressed chemokine (TECK) have similar chemoattractive capacities on the same thymocyte populations as does ABCD-1 (Campbell et al., 1999), suggesting that they may substitute for the loss of ABCD-1. The role(s) for ABCD-2 in the thymus still remain(s) to be elucidated, since neither $CD4^+CD8^-$, $CD4^-CD8^+$, $CD69^+$ nor $CD69^-$ thymocytes responded to it (Lieberam and Förster, 1999).

The targets for ABCD-1 and -2 chemoattraction are activated T cells. In mice, we have seen no significant difference in the migratory behavior of either $CD4^+$ or $CD8^+$ T lymphoblasts toward ABCD-1 or -2. Murine or human T cells polarized *in vitro* towards Th1 or Th2 both were equally attracted by ABCD-1 or -2 (Schaniel et al., 1998; 1999). In contrast, human ABCD-1 and -2 have been reported to chemoattract preferentially human Th2, but not Th1 cells (Andrew et al., 1998; Bonecchi et al., 1998; Imai et al., 1999). This discrepancy might be explained by the different activation and polarization protocols used in these studies.

ABCD-3 consists of a chemokine-like domain, followed by a mucin-like stalk, and a transmembrane region. Different migratory activities were observed *in vitro* and *in vivo* depending on which part of the ABCD-3 molecule was used in the chemoattractive experiments (Bazan et al., 1997; Pan et al., 1997). Bazan and colleagues showed migration of peripheral blood monocytes, to a lesser extent of T cells, but not of neutrophils when the entire extracellular portion of human ABCD-3 was used as an attractant (Bazan et al., 1997). In their analysis, the chemokine-like domain plus half of the mucin-like stalk as well as a synthetic chemokine-only domain protein proved to be as potent as the entire extracellular portion of ABCD-3 for the T lymphocytes, but not for the monocytes (Bazan et al., 1997). The chemokine-like domain used by Pan and colleagues was chemoattractive for T cells and neutrophils, but not monocytes *in vitro*. However, it was able to recruit T lymphocytes and neutrophils, and also monocytes into the peritoneum of mice (Pan et al., 1997). The authors think this to be a secondary effect seen *in vivo* which does not reflect the chemotactic specificity of ABCD-3. In the same study, the extracellular portion of human ABCD-3 failed to attract monocytes and T lymphocytes *in vivo* which was consistent with their *in vitro* data (Pan et al., 1997). However, in contrast to their observation *in vitro*, the extracellular portion of ABCD-3 was chemotactic for neutrophils, suggesting to them differential activation of ABCD-3 *in vivo*. In our hands, murine ABCD-3, when compared to ABCD-1 and -2, did not significantly induce migration of either peripheral $CD4^+$ or $CD8^+$ T cells, or CD4 T lymphocytes polarized towards Th1 or Th2 (Schaniel et al., 1999). In fact, it is presently unknown which cells of the mouse are attracted by ABCD-3.

ABCD-3, a special chemokine containing a mucin-like structure and a transmembrane region, might play a role as adhesive molecule in leukocyte capture and firm adhesion. In fact, this has been demonstrated for resting monocytes, resting and IL-2 activated $CD8^+$ T cells and IL-2 stimulated NK cells, but not for resting or IL-2 activated $CD4^+$ T cells (Fong et al., 1998; Imai et al., 1997).

Receptors Recognizing the ABCD Chemokines

The chemokine receptor CCR4 was identified as a highly specific receptor for both ABCD-1 and -2 (Imai et al., 1997; Imai et al., 1998). CCR4, originally isolated from a human basophilic cell line, was strongly expressed in thymus, peripheral blood T cells but not in B cells, NK cells, monocytes and granulocytes (Imai et al., 1997). Expression analysis of CCR4 on polarized T cells revealed that CCR4 production was associated preferentially with Th2 cells (Bonecchi et al., 1998; Sallusto et al., 1998a). Our own data on the desensitization of Ca^{++} fluxes and on cellular migration of activated T cells, Th1 as well as Th2 cells, suggest that ABCD-1 recognizes not only CCR4 but also another, yet unidentified receptor. This unknown receptor, in contrast to CCR4, is not recognized by ABCD-2 (Schaniel et al., 1999).

ABCD-3 was found to be a functional ligand for the orphan receptor V28, now termed CX_3CR1 (Imai et al., 1997). CX_3CR1 was found expressed on human T cells, NK cells and monocytes (for references see Imai et al., 1997), and mediated both leukocyte migration and adhesion.

A Model for the Roles of the ABCD Chemokines in Cell Migration in Secondary Lymphoid Organs, in T Cell-dependent B Cell Responses and Germinal Center Formation

The distinct migration into and within secondary lymphoid organs, spleen, LN, tonsils and PP, and the cooperation of many cells of the lymphoid system are key requirements for a proper immune response against any microbe or other antigen. The secondary lymphoid organs are sites where specialized cells harbor and concentrate antigen or its processed peptides. Blood-borne antigens are filtered and concentrated within the spleen, antigens entering through the skin are carried by the lymphatic system into draining LN, and those antigens invading the gastrointestinal tract are carried by specialized epithelial cells into PP and tonsils. The secondary lymphoid organs are also the sites where interactions of DC presenting antigenic-peptides in the context of MHC class II with T cells and of B cells with T cells take place leading to the formation of primary follicles and, in the end, of germinal centers (Melchers et al., 1999).

The first cells normally involved in an immune response are immature DC. When immature DC encounter antigen in the infected peripheral tissue and are stimulated by inflammatory cytokines they get activated and mature to produce high amounts of inflammatory chemokines, thereby recruiting more immature DC, macrophages and effector T cells to the site of inflammation. This is a critical event for the initiation of the immune response. Maturing DC also migrate from the infected peripheral tissue into the paracortex of LN or the periarteriolar lymphoid sheath of spleen where they could present the antigenic peptides in the context of MHC to naïve T cells (Fig. 1). As part of their maturation process the DC down-regulate expression of inflammatory chemokine receptors, up-regulate expression of CCR7 and become responsive to SLC and ELC (Sallusto et al., 1998b). SLC, in addition to being expressed by stromal cells within secondary lymphoid tissues, is produced by lymphatic endothelium, including high endothelium venules (Gunn et al., 1998a), thus likely attracting the DC into the tissue (Fig. 1). ELC expressed by resident, mature DC within the T cell zone of secondary lymphoid organs then homes the migrating DC into the T cell areas. Migration of naïve T cells to secondary lymphoid organs is also regulated by CCR7 and its ligand SLC (Förster et al., 1999).

Once the virgin T cells have reached the secondary lymphoid tissue they have to be scanned for recognition of antigen presented by the DC. ELC, and in humans also dendritic cell chemokine (DC-CK)-1 (Adema et al., 1997), produced by DC in the T cell areas likely regulate this step. The naïve T cell is activated by interaction with the DC presenting the TCR-fitting antigenic peptide. Expression of ABCD-1 and -2 by the DC could now keep the primed T cell and its daughters attracted to it or to newly arriving DC, and, hence, continuously activated (Fig. 1). This process likely accelerates the rate of encounter between antigen-presenting DC and antigen-specific T cells, leading to expansion of the antigen-specific T cells.

Furthermore, the production of the transmembrane chemokine ABCD-3 by the DC may suggest a function in DC-to-T cell adhesion. B cells in the mantle zone and the follicles also have to be activated. Presumably ELC, and maybe also SLC, then attract the activated B cells toward the DC associated with activated antigen-specific T cells. BCR stimulation promotes B cells to produce moderate levels of ABCD-1 and -2 that might enhance the chance of encounter between antigen-presenting B cells and antigen-specific T cells, allowing interactions through CD40-CD40L. At the same time, the B cells are attracted into the follicles by B-lymphocyte chemoattractant (BLC) which is probably produced by follicular dendritic cells (FDC) (Gunn et al., 1998b; Legler et al., 1998). The chemokines ABCD-1 and -2 produced by the activated B cells, and maybe also BLC produced by the FDC (Ansel et al., 1999), support recruitment of the activated antigen-specific T cells into the follicles allowing specific cooperation of the T cells with the B cells, leading to the formation of a germinal center (Fig. 1).

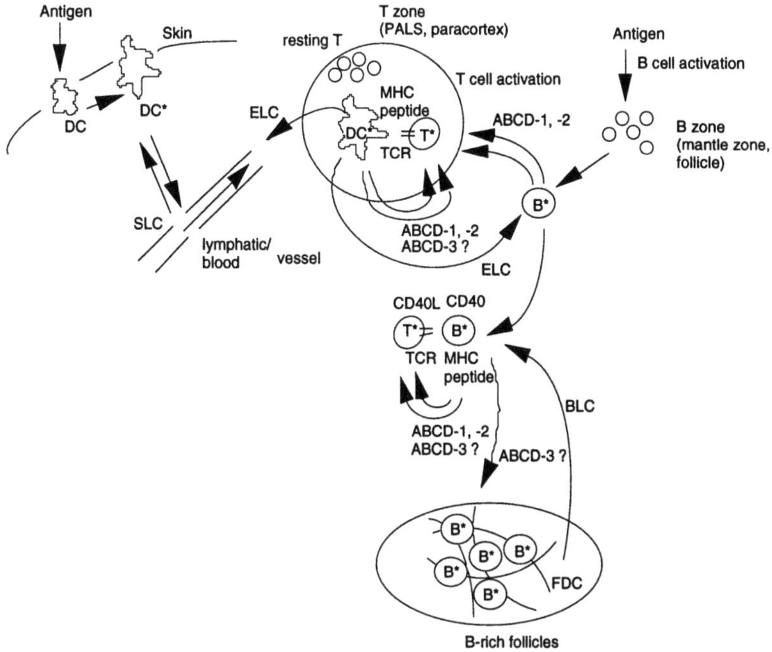

Fig. 1. The chemokines SLC, ELC, ABCD-1, -2 and -3, and BLC orchestrate a T cell-dependent immune response of B cells in a secondary lymphoid organ.
DC = dendritic cell; FDC = follicular dendritic cell; * = activated cell. For details see text.

Acknowledgements

The Basel Institute for Immunology was founded and is supported by F. Hoffmann-La Roche Ltd., Basel, Switzerland.

References

Adema GJ, Hartgers F, Verstraten R, de Vries E, Marland G, Menon S, Foster J, Xu Y, Nooyen P, McClanahan T, Bacon KB, Figdor CG (1997) A dendritic-cell-derived C-C chemokine that preferentially attracts naive T cells. Nature 387: 713-717

Andrew DP, Chang MS, McNinch J, Wathen ST, Rihanek M, Tseng J, Spellberg JP,Elias III CG (1998) STCP-1 (MDC) CC chemokine acts specifically on chronically activated Th2 lymphocytes and is produced by monocytes on stimulation with Th2 cytokines IL-4 and IL-13. J Immunol 161: 5027-5038

Ansel K M, McHeyzer-Williams LJ, Ngo VN, McHeyzer-Williams MG, Cyster JG (1999) In vivo-activated CD4 T cells upregulate CXC chemokine receptor 5 and reprogram their response to lymphoid chemokines. J Exp Med 190: 1123-1134

Bazan JF, Bacon KB, Hardiman G, Wang W, Soo K, Rossi D, Greaves DR, Zlotnik A, Schall TJ (1997) A new class of membrane-bound chemokine with a CX3C motif. Nature 385: 640-644

Bonecchi R, Bianchi G, Bordignon PP, D'Ambrosio D, Lang R, Borsatti A, Sozzani S, Allavena P, Gray PA, Mantovani A, Sinigaglia F (1998) Differential expression of chemokine receptors and chemotactic responsiveness of type 1 T helper cells (Th1s) and Th2s. J Exp Med 187: 129-134

Butcher EC, Williams M, Youngman K, Rott L, Briskin M (1999) Lymphocyte trafficking and regional immunity. Adv Immunol 72: 209-253

Campbell JJ, Pan J, Butcher EC (1999) Cutting edge: developmental switches in chemokine responses during T cell maturation. J Immunol 163: 2353-2357

Chang M, McNinch J, Elias III C, Manthey CL, Grosshans D, Meng T, Boone T, Andrew DP (1997) Molecular cloning and functional characterization of a novel CC chemokine, stimulated T cell chemotactic protein (STCP-1) that specifically acts on activated T lymphocytes. J Biol Chem 272: 25229-25237

Chantry D, DeMaggio AJ, Brammer H, Raport J, Wood CL, Schweickart VL, Epp A, Smith A, Stine JT, Walton K, Tjoelker L, Godiska R, Gray PW (1998) Profile of human macrophage transcripts: insights into macrophage biology and identification of novel chemokines. J Leukoc Biol 64: 49-54

Fong AM, Robinson LA, Steeber DA, Tedder TF, Yoshie O, Imai T, Patel DD (1998) Fractalkine and CX3CR1 mediate a novel mechanism of leukocyte capture, firm adhesion, and activation under physiologic flow. J Exp Med 188: 1413-1419

Förster R, Schubel A, Breitfeld D, Kremmer E, Renner-Müller I, Wolf E, Lipp M (1999) CCR7 coordinates the primary immune response by establishing functional microenvironments in secondary lymphoid organs. Cell 99: 23-33

Galli G, Chantry D, Annunziato F, Romagnani P, Cosmi L, Lazzeri E, Manetti R, Maggi E, Gray PW, Romagnani S (2000) Macrophage-derived chemokine production by activated human T cells in vitro and in vivo: preferential association with the production of type 2 cytokines. Eur J Immunol 30: 204-210

Godiska R, Chantry D, Raport CJ, Sozzani S, Allavena P, Leviten D, Mantovani A, Gray PW (1997) Human macrophage-derived chemokine (MDC), a novel chemoattractant for monocytes, monocyte-derived dendritic cells, and natural killer cells. J Exp Med 185: 1595-1604

Gonzalo JA, Pan Y, Lloyd CM, Jia GQ, Yu G, Dussault B, Powers CA, Proudfoot AE, Coyle AJ, Gearing DGutierrez-Ramos JC (1999) Mouse monocyte-derived chemokine is involved in airway hyperreactivity and lung inflammation. J Immunol 163: 403-411

Gunn MD, Tangemann K, Tam C, Cyster JG, Rosen SD, Williams LT (1998a) A chemokine expressed in lymphoid high endothelial venules promotes the adhesion and chemotaxis of naive T lymphocytes. Proc Natl Acad Sci U S A 95: 258-263

Gunn MD, Ngo VN, Ansel KM, Ekland EH, Cyster JG, Williams LT (1998b) A B-cell-homing chemokine made in lymphoid follicles activates Burkitt's lymphoma receptor-1. Nature 391: 799-803

Harrison JK, Jiang Y, Chen S, Xia Y, Maciejewski D, McNamara RK, Streit WJ, Salafranca MN, Adhikari S, Thompson DA, Botti P, Bacon KB, Feng L (1998) Role for neuronally derived fractalkine in mediating interactions between neurons and CX3CR1-expressing microglia. Proc Natl Acad Sci U S A 95: 10896-10901

Imai T, Baba M, Nishimura M, Kakizaki M, Takagi S, Yoshie O (1997) The T cell-directed CC chemokine TARC is a highly specific biological ligand for CC chemokine receptor 4. J Biol Chem 272: 15036-15042

Imai T, Chantry D, Raport CJ, Wood CL, Nishimura M, Godiska R, Yoshie O, Gray PW (1998) Macrophage-derived chemokine is a functional ligand for the CC chemokine receptor 4. J Biol Chem 273: 1764-1768

Imai T, Hieshima K, Haskell C, Baba M, Nagira M, Nishimura M, Kakizaki M, Takagi S, Nomiyama H, Schall TJ, Yoshie O (1997) Identification and molecular characterization of fractalkine receptor CX3CR1, which mediates both leukocyte migration and adhesion. Cell 91: 521-530

Imai T, Nagira M, Takagi S, Kakizaki M, Nishimura M, Wang J, Gray PW, Matsushima K, Yoshie O (1999) Selective recruitment of CCR4-bearing Th2 cells toward antigen-presenting

cells by the CC chemokines thymus and activation-regulated chemokine and macrophage-derived chemokine. Int Immunol 11: 81-88
Imai T, Yoshida T, Baba M, Nishimura M, Kakizaki M, Yoshie O (1996) Molecular cloning of a novel T cell-directed CC chemokine expressed in thymus by signal sequence trap using Epstein-Barr virus vector. J Biol Chem 271: 21514-21521
Kanazawa N, Nakamura T, Tashiro K, Muramatsu M, Morita K, Yoneda K, Inaba K, Imamura S, Honjo T (1999) Fractalkine and macrophage-derived chemokine: T cell-attracting chemokines expressed in T cell area dendritic cells. Eur J Immunol 29: 1925-1932
Legler DF, Loetscher M, Roos S, Clark-Lewis I, Baggiolini M, Moser B (1998) B cell-attracting chemokine 1, a human CXC chemokine expressed in lymphoid tissues, selectively attracts B lymphocytes via BLR1/CXCR5. J Exp Med 187: 655-660
Lieberam I, Förster I (1999) The murine beta-chemokine TARC is expressed by subsets of dendritic cells and attracts primed CD4+ T cells. Eur J Immunol 29: 2684-2694
Lloyd CM, Delaney T, Nguyen T, Tian J, Martinez AC, Coyle AJ, Gutierrez-Ramos JC (2000) CC chemokine receptor (CCR)3/Eotaxin is followed by CCR4/Monocyte- derived chemokine in mediating pulmonary T helper lymphocyte type 2 recruitment after serial antigen challenge in vivo. J Exp Med 191: 265-274
Maciejewski-Lenoir D, Chen S, Feng L, Maki R, Bacon KB (1999) Characterization of fractalkine in rat brain cells: migratory and activation signals for CX3CR-1-expressing microglia. J Immunol 163: 1628-1635
Melchers F, Rolink AG, Schaniel C (1999) The role of chemokines in regulating cell migration during humoral immune responses. Cell 99: 351-354
Nomiyama H, Imai T, Kusuda J, Miura R, Callen DF, Yoshie O (1997) Assignment of the human CC chemokine gene TARC (SCYA17) to chromosome 16q13. Genomics 40: 211-213
Pan Y, Lloyd C, Zhou H, Dolich S, Deeds J, Gonzalo JA, Vath J, Gosselin M, Ma J, Dussault B, Woolf E, Alperin G, Culpepper J, Gutierrez-Ramos JC, Gearing D (1997) Neurotactin, a membrane-anchored chemokine upregulated in brain inflammation. Nature 387: 611-617
Rodenburg RJ, Brinkhuis RF, Peek R, Westphal JR, Van Den Hoogen FH, van Venrooij WJ, van de Putte LB (1998) Expression of macrophage-derived chemokine (MDC) mRNA in macrophages is enhanced by interleukin-1beta, tumor necrosis factor alpha, and lipopolysaccharide. J Leukoc Biol 63: 606-611
Ross R, Ross XL, Ghadially H, Lahr T, Schwing J, Knop J, Reske-Kunz AB (1999) Mouse langerhans cells differentially express an activated T cell- attracting CC chemokine. J Invest Dermatol 113: 991-998
Rossi DL, Hardiman G, Copeland NG, Gilbert DJ, Jenkins N, Zlotnik A, Bazan JF (1998) Cloning and characterization of a new type of mouse chemokine. Genomics 47: 163-170
Sallusto F, Lenig D, Mackay CR, Lanzavecchia A (1998a) Flexible programs of chemokine receptor expression on human polarized T helper 1 and 2 lymphocytes. J Exp Med 187: 875-883
Sallusto F, Schaerli P, Loetscher P, Schaniel C, Lenig D, Mackay CR, Qin S, Lanzavecchia A (1998b) Rapid and coordinated switch in chemokine receptor expression during dendritic cell maturation. Eur J Immunol 28: 2760-2769
Schaniel C, Pardali E, Sallusto F, Speletas M, Ruedl C, Shimizu T, Seidl T, Andersson J, Melchers F, Rolink AG, Sideras P (1998) Activated murine B lymphocytes and dendritic cells produce a novel CC chemokine which acts selectively on activated T cells. J Exp Med 188: 451-463
Schaniel C, Sallusto F, Ruedl C, Sideras P, Melchers F, Rolink AG (1999) Three chemokines with potential functions in T lymphocyte-independent and -dependent B lymphocyte stimulation. Eur J Immunol 29: 2934-2947
Springer TA (1994) Traffic signals for lymphocyte recirculation and leukocyte emigration: the multistep paradigm. Cell 76: 301-314
Tang HL, Cyster JG (1999) Chemokine up-regulation and activated T cell attraction by maturing dendritic cells. Science 284: 819-822
Zlotnik A, Morales J, Hedrick JA (1999) Recent advances in chemokines and chemokine receptors. Crit Rev Immunol 19: 1-47

Affinity Maturation in Ectopic Germinal Centers

B.A. de Boer[1], I. Voigt[1], H.-J. Kim[1], S.A. Camacho[1,3], M. Lipp[2], R. Förster[2] and C. Berek[1]

[1]Deutsches Rheuma ForschungsZentrum, Monbijoustr. 2, 10117 Berlin, Germany

[2]Max-Dellbrück-Center for Molecular Medicine, Berlin-Buch, Germany

[3]present address: The Scripps Research Institute, 10550 North Torrey Pines Rd, La Jolla, USA

Introduction

It is becoming clear that a precisely structured lymphoid tissue is essential for the establishment of an antigen specific immune response. The differentiation of the antigen activated lymphocyte into effector cells requires a well organized microenvironment in which the antigen specific and the antigen presenting cells come into close physical contact. The central role of cytokines in lymphoid organogenesis and the important function of chemokines for guided cell movement into and within the secondary lymphoid organs are just emerging.

Organization of the Splenic Tissue

T and B cells are well separated into distinct areas within the white pulp of the spleen. T cells are mainly found in a periarteriolar lymphoid sheath surrounding the central arterioles, whereas B cells are concentrated in eccentric cuffs, the primary follicles. The basis of the primary follicle is formed by a network of follicular dendritic cells (FDC).

Genetically manipulated mice have demonstrated the essential function of tumor necrosis factor α (TNFα) and lymphotoxin α (LTα) for the organization of the lymphoid tissue [1]. Mice deficient for these proinflammatory cytokines or for their receptors lack the network of FDC and fail to form primary B cell follicles.

Chemokines are important for the homing of the T and B cells into their compartments. Repopulation experiments showed that B cells deficient for the chemokine CXC receptor 5 (CXCR5) fail to find their way into the primary follicle [2]. This receptor interacts with the B cell attracting chemokine (BCA-1) also known as B lymphocyte chemoattractant (BLC) [3,4]. In situ hybridization demonstrated that this chemokine is constitutively expressed in the primary B cell follicles. One interpretation of these results is that FDC secrete BLC / BCA and in this way attract B cells into the follicle [3].

Germinal Center Reaction

B cell activation is a complex process which requires an intimate association of B cells, T cells and antigen presenting cells in the T cell zone. As a result, the antigen activated B cell starts to proliferate and the developing clone grows into the network of FDC. Within a few days large clones are generated which displace the naive B cells from the primary follicle [5,6]. The primary follicle has developed into a germinal center (GC).

Only in the second week after immunization does the classical structure of the GC with a dark and a light zone emerge [6]. By this time, only a few dividing centroblasts are still in the network of FDC and proliferation is largely restricted to the dark zone. In the dividing B cells hypermutation is activated and nucleotide exchanges are introduced into the rearranged V-region genes generating affinity variants [7,8]. Upon contact with antigen presented by the FDC of the light zone the few cells with relatively high affinity receptors are selected to differentiate into plasma and memory cells.

The Organization of the Splenic Tissue in the CXCR5-Deficient Mouse

In mice deficient for the chemokine CXCR5 cells cannot respond to the chemokine BCA / BLC. This interferes with their localization to the primary follicles and as a result polarized primary B cell follicles are not formed in the spleen [2]. Nevertheless, labeling for the M2 determinant on FDC demonstrated that in CXCR5-deficient mice FDC and B cells do colocalize. Both cell types are found along the marginal sinuses (Fig. 1, left). This unexpected result stresses how little we know about the segregation of T and B cells and the underlying mechanisms allowing the establishment of the primary B cell follicle.

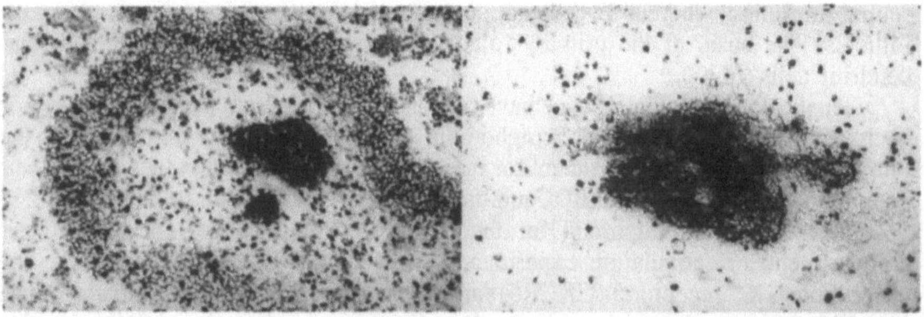

Fig. 1. Ectopic GC. A GC from the splenic tissue of a CXCR5-deficient mouse (left) and from the synovial tissue of a patient with rheumatoid arthritis (RA) (right) are shown. In both tissues, proliferating B cells expressing the nuclear antigen Ki-67 are embedded within the network of FDC. Double labeling with affinity purified rabbit antibodies specific for the nuclear antigen Ki-67 (Dako) and with a mAb specific for FDC (mAb M2 [9] and Wue-2 [10], specific for murine or human FDC respectively) was performed. Original magnification x 100.

When CXCR5-deficient mice are immunized with the T cell dependent antigen 2-phenyl-oxazolone (phOx) small clusters of B cells become apparent in the T cell zone. These antigen activated B cells proliferate in a network of FDC surrounding the central arteriole (Fig. 1, left). Dividing B cells showed strong binding to the lectin peanut agglutinin, a marker for GC B cells [11], indicating that in these CXCR5-deficient mice ectopic GC form in the T cell zone. In CXCR5-deficient animals the classical structure of the GC with a dark and a light zone does not develop, even two weeks after immunization, when large GC of several 1000 cells are generated. The proliferating B cells are still embedded within the network of FDC (Fig. 1, left).

Affinity Maturation in Ectopic GC

To study the process of affinity maturation in the ectopic GC, CXCR5-deficient mice were immunized with phOx and two weeks later GC B cells were isolated by micromanipulation direct from frozen splenic tissue sections [12]. DNA was extracted and amplified with a panel of primers, specific for the mouse Vκ L-chains. This analysis revealed that an oligoclonal B cell population is expanded in these GC. As described for the normal BALB/c mouse, the immune response to phOx is dominated by B cells expressing a VκOx1 L-chain. [15].

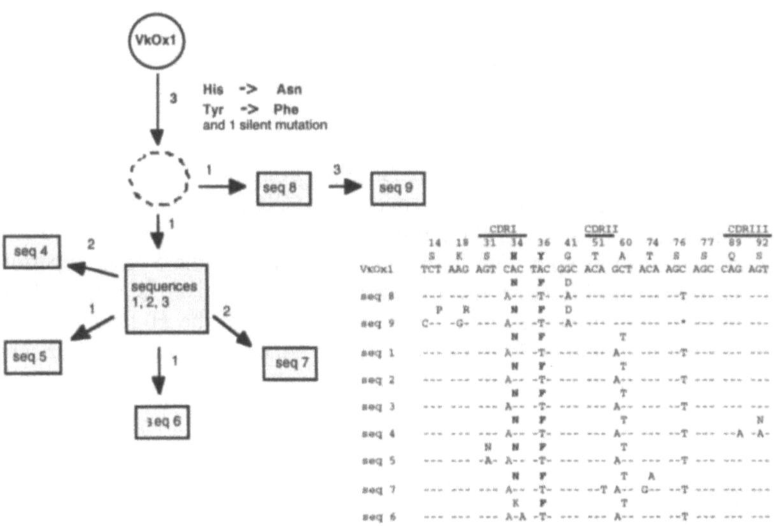

Fig. 2. Affinity maturation in an ectopic GC. The genealogical tree of the intraclonal diversification is shown. The numbers besides the arrows refer to the numbers of nucleotide exchanges that distinguish one sequence from the other. VκOx1 sequences were compared to the germline V-gene [13]. Only nucleotide differences are shown. CDRs are indicated, numbering according to Kabat [14].

Sequences with identical VκOx1 to Jκ5 rearrangement but a different pattern of somatic mutations were isolated from a single GC (Fig. 2). The genealogical tree shows the stepwise accumulation of somatic mutations. These results demonstrate that during proliferation in the network of FDC the hypermutation is activated.

Sequences isolated from this B cell clone had three somatic mutations in common. Besides a silent mutation at position 76, all sequences carried 2 replacement mutations which are characteristic for the phOx specific antibodies of high affinity. The data suggest that from the broad spectrum of sequences, generated by hypermutation, those rare variants expressing somatic mutations known to increase the affinity for the antigen phOx are selected.

The analysis of the immune response to phOx suggests that an ectopic GC developing in the T cell zone supports the process of affinity maturation through hypermutation and selection. This ectopic structure functions efficiently since the selection of high affinity variants is as rapid as in a normal GC. The data demonstrate that a compartmentalization of the GC into a dark and a light zone is not a prerequisite for affinity maturation. The network of FDC can support proliferation and activation of the hypermutation mechanism as well as selection of the high affinity variants.

Ectopic GC in Autoimmune Diseases

In autoimmune diseases tertiary lymphoid tissue may develop in the affected tissue. During chronic inflammation high levels of the proinflammatory cytokines, like TNFα and LTα, may support lymphoid organogenesis. In patients with RA well organized follicle-like structures develop in the inflamed synovium [16]. In these lymphoid structures a clear B / T cell segregation is missing. Nevertheless, in many cases the center of these structures is practically free of T cells. Here, a perivascular network of FDC is established in which B cells proliferate (Fig. 1, right). As described for the ectopic GC in the T cell zone of the CXCR5-deficient mice B cells in the FDC network express a GC phenotype. They are positive for CD38 and have downregulated their Ig receptor. Furthermore, the analysis of the V-gene repertoire revealed that during proliferation the hypermutation mechanism is activated and somatic mutations are introduced into the V-region genes of the H- and L-chains. Again, the classical structure of the GC is missing, since the proliferating B cells are within the network of FDC (Fig. 1, right).

Conclusions

The analysis of the CXCR5-deficient mouse has demonstrated that the process of affinity maturation may take place in the ectopic GC. A strict compartmentalization of the GC into a dark and a light zone is not a necessary prerequisite for affinity selection. Similarly, in autoimmune diseases in which ectopic GC develop in chronically inflamed tissue, the FDC network may support both B cell activation and differentiation into high affinity effector cells. It seems that in organ-specific autoimmune diseases like myasthenia gravis or Graves′ disease, the autoreactive antibodies of high affinity are generated within the affected tissues [17]. In other immune diseases, such as RA, it remains to be shown whether the B cell response in the affected tissue is part of the pathogenetic processes. It is well possible that these ectopic GC function as an additional tertiary lymphoid tissue. In such cases activation of the B cells in the synovial tissue may be merely a bystander effect of an immune response which would normally be established in the peripheral lymphoid organs.

Acknowledgements
This work was supported by the DFG grant Be-1171/1-3 to CB. The DRFZ is supported by the Berlin Senate of Research and Education.

References

1. Fu Y-X, Chaplin DD (1999) Development and maturation of secondary lymphoid tissue. Ann Rev Immunol 17:399-433
2. Förster R, Mattis AE, Kremmer E, Wolf E, Brem G, Lipp M (1996) A putative chemokine receptor, BLR1, directs B cell migration to defined lymphoid organs and specific anatomic compartments of the spleen. Cell 87:1037-1047
3. Gunn DM, Ngo VN, Ansel KM, Ekland EH, Cyster JG, Williams LT (1998) A B-cell-homing chemokine made in lymphoid follicles activates Burkitt′s lymphoma receptor-1. Nature 391:799-803
4. Legler DF, Loetscher M, Roos RS, Clark-Lewis I, Baggiolini M, Moserr B (1998) B cell-attracting chemokine 1, a human CXC chemokine expressed in lymphoid tissues, selectively attracts B lymphocytes via BLR1/CxCR5. J Exp Med 187:655-660
5. MacLennan ICM (1994) Germinal centers. Ann Rev Immunol 12:117-139
6. Camacho SA, Kosco-Vilbois MH, Berek C (1998) The dynamic structure of the germinal center. Immunol Today 19:511-514
7. Berek C, Berger A, Apel M (1991) Maturation of the immune response in germinal centers. Cell 67:1121-1129
8. Jacob J, Kelsoe G, Rajewsky K, Weiss U (1991) Intraclonal generation of antibody mutants in germinal centres. Nature 354:389-392
9. Kosco-Vilbois MH, Zentgraf H, Gerdes J, Bonnefoy J-Y (1997) To "B" or not to "B" a germinal center? Immunol Today 18:225-230
10. Schröder AE, Greiner A, Seyfert C, Berek C (1996) Differentiation of B cells in the non-lymphoid tissue of the synovial membrane of patients with rheumatoid arthritis. Proc Natl Acad Sci USA 93:221-225
11. Rose ML, Birbeck MC, Wallis VJ, Forrester JA, Davies AJS (1980) Peanut lectin binding properties of germinal centres of mouse lymphoid tissue. Nature 284:364-366
12. Voigt I, Camacho SA, de Boer BA, Lipp M, Förster R, Berek C (2000) CXCR5-deficient mice develop functional germinal centers in the splenic T cell zone. Eur J Immunol 30:560-567
13. Even J, Griffiths GM, Berek C, Milstein C (1985) Light chain germ-line genes and the immune response to 2-phenyloxazolone. EMBO J 4:3439-3445
14. Kabat EA, Wu TT, Perry HM, Gottesman KS, Foeller C (1991) Sequences of proteins of immunological interest. NIH Publication,Bethesda,Maryland,USA 2:91-3242
15. Berek C, Milstein C (1987) Mutation drift and repertoire shift in the maturation of the immune response. Immunol Rev 96:23-41
16. Kim H-J, Krenn V, Steinhauser G, Berek C (1999) Plasma cell development in synovial germinal centers in patients with rheumatoid and reactive arthritis. J Immunol 162:3053-3062
17. Chazenbalk GD, Protolano S, Russo D, Hutchinson JS, Rapoport B, McLachlan S (1993) Human organ-specific autoimmune disease. Molecular cloning and expression of an autoantibody gene repertoire for a major autoantigen reveals an antigenic immunodominant region and restricted immunoglobulin gene usage in the target organ. J Clin Invest 92:62-74

Precursors to Neonatal Lymph Nodes: $LT\beta^{+}CD45^{+}CD4^{+}CD3^{-}$ Cells are Found in Fetal Liver

[1]R. Mebius and [2]K. Akashi

[1]Department of Cell Biology and Immunology, Faculty of Medicine, Vrije Universiteit, 1081 BT Amsterdam, The Netherlands
[2]Departments of Pathology and Developmental Biology, Stanford University School of Medicine, Stanford, California 94305, USA

Lymphoid Organ Development

Lymph nodes are strategically positioned throughout the body. All lymph, draining from tissues and organs, passes through lymph nodes before entering the bloodstream, which allows for efficient screening of incoming pathogens or tissue aberrations. It is now well established that the highly organized structure of lymph nodes in various compartments is essential for an optimal interaction of lymphocytes and antigen presenting cells to induce immune responses.

The development of lymphoid organs and the organization within these organs is established during ontogeny, while the maintenance of this organization within lymphoid organs requires signals continuously throughout life (Fu and Chaplin, 1999). During ontogeny, the forming of lymph nodes, and also of Peyer's patches, is a highly coordinated process involving selective and transient expression of a number of obligatory genes. For the proper development of lymph nodes and Peyer's patches, lymphotoxin-α (LTα), lymphotoxin-β (LTβ), and lymphotoxin-β receptor (LTβ-R) are required and it was shown that signalling through the LTβ-R by cell surface expressed LTαβ is crucial (Alimzhanov et al., 1997; Banks et al., 1995; Futterer et al., 1998; Rennert et al., 1996; Rennert et al., 1998; Togni et al., 1994). In addition, it is now clear that many other genes, including those encoding IL-7, IL-7R, relB, IKAROS, CXCR5, Id-2, and OPGL/TRANCE are involved in lymphoid organ development (Burkly et al., 1995; Cao et al., 1995; Forster et al., 1996; Freeden-Jeffry et al., 1995; Georgopoulos et al., 1994; Kong et al., 1998; Wang et al., 1996; Weih et al., 1995; Yokota et al., 1999). However, the cellular requirements and sequence of events involving these gene products are unknown.

The most widely accepted view of lymph node development was originally postulated by Sabin (Sabin, 1909). He proposed that the earliest event in lymph node development involves the budding of lymph sacs from the larger veins (Sabin, 1909). These lymph sacs eventually break off from the larger veins. Then, the lymphatic vascular system develops by endothelial sprouting from these lymph sacs. Invaginations of connective tissue push into the lumen of these sacs, forming the anlage of the lymph nodes. In mice the earliest lymph sacs can be seen at E12.5 by specific expression of fms-like tyrosine kinase 4 (FLT4) and Prox-1, two genes that show restricted expression in a subset of endothelial cells and lymphatic vessels (Kaipainen et al., 1995; Wigle and Oliver, 1999). The connective tissue invaginations that subsequently protrude into these sacs, ultimately forming the lymph nodes, push the lumen of the lymphatic sacs to the side. These compressed sacs become the subcapsular sinuses of lymph nodes. Before birth these developing lymph nodes are colonized selectively by TCR$\gamma\delta$ T cells and $CD45^+CD4^+CD3^-$ cells in a MAdCAM-1/$\alpha4\beta7$ dependent manner (Mebius et al., 1996). Since cell surface expression of lymphotoxin has been shown to be crucial for lymph node development, our observation that $Lt\alpha_1\beta_2$ is expressed by $\alpha4\beta7^+CD45^+CD4^+CD3^-CD62L^-$ cells, makes these cells likely candidates to deliver some of the obligatory signals for lymph node induction (Mebius et al., 1997). $CD45^+CD4^+CD3^-$ cells have been described to be able to give rise to NK cells, antigen presenting cells, when placed in IL-2, or IL-4 and GM-CSF, respectively. Also, they do not belong to the rearranging subset of lymphocytes (Mebius et al., 1997). *In vivo*, these cells specifically tend to home to B-cell follicles (Mebius et al., 1997).

Fetal Liver Precursors

In order to unravel the precise mechanisms of lymphoid organ development, we characterized the precursors to $CD45^+CD4^+CD3^-$ cells (Mebius et al., submitted for publication). Hereto fetal liver subsets were analyzed for the presence of precursor populations with a similar phenotype as the $CD45^+CD4^+CD3^-$ cells. First, expression of CD4 in fetal liver cells was determined and no distinct population with sufficient high expression levels could be observed (Mebius et al., 1997). Another cell surface marker expressed by $CD45^+CD4^+CD3^-$ cells and associated with a precursor phenotype is expression of the interleukin-7 receptor (IL7R), while low levels of c-kit could be observed (Mebius et al., 1997). In adult bone marrow the existence of lymphoid restricted progenitors, the common lymphoid progenitors (CLP) have been described (Kondo et al., 1997). These cells are clonogenic precursors to B lymphocytes and T lymphocytes, and additionally can give rise to NK cells. The CLP in adult bone marrow show high expression levels of IL7R and moderate levels of c-Kit and Sca-1 (Kondo et al., 1997). We therefore analyzed fetal livers for the presence of a subset with similarities to the CLP population in adult bone marrow. Such a subset could be found, and this subset was further analyzed for different cell lineage markers. Analysis of CD4, CD8, Gr-1 expression showed that these antigens were all absent on the $IL7R^+c\text{-}Kit^{lo}Sca\text{-}1^{lo}$ cells in fetal liver, while B220 and Mac-1 showed low to no expression. Thus, $IL7R^+c\text{-}Kit^{lo}Sca\text{-}1^{lo}$ cells could be found in fetal livers and this population forms the fetal liver counterpart to adult CLP (Kondo et al., 1997; Mebius et al., submitted for publication).

Differentiation Capacity of Fetal Liver $IL7R^+c\text{-}Kit^{lo}Sca\text{-}1^{lo}$ Cells *in vivo*

In order to test the potential of the fetal liver precursors to give rise to various hematopoietic lineages, these cells were injected into sublethally irradiated hosts. To distinguish donor from host cells we used the congenic mouse strains C57BL/Ka-Ly5.1 and C57BL/Ka-Ly5.2, which differ only at the Ly5 locus, allowing for detection of donor derived cells.

Analysis of mice at several timepoints after injection revealed that the injected fetal liver cells, characterized by expression of IL7R and low levels of c-Kit and

Sca-1, predominantly gave rise *in vivo* to B lymphocytes and T lymphocytes. In addition, we also observed NK cells and CD8α^+ and CD8α^- dendritic cells as progeny from the injected cells. Detailed analysis of lymph nodes and spleen showed that CD45$^+$CD4$^+$CD3$^-$ cells were also derived from IL7R$^+$c-Kit^{lo}Sca-1^{lo} cells. However, no macrophages or granulocytes as progeny from the IL7R$^+$c-Kit^{lo}Sca-1^{lo} cells could be found *in vivo*.

Differentiation Capacity of Fetal Liver IL7R$^+$c-Kit^{lo}Sca-1^{lo} Cells *in vitro*
To see if the differentiation potential of the IL7R$^+$c-Kit^{lo}Sca-1^{lo} cells as seen *in vivo* was reflected *in vitro*, we analyzed the differentiation potential of IL7R$^+$c-Kit^{lo}Sca-1^{lo} fetal liver cells on bone marrow stromal cells in combination with cytokines that allow B-lymphocyte and myeloid differentiation. In these assays we observed differentiation of IL7R$^+$c-Kit^{lo}Sca-1^{lo} cells towards the B-lymphocyte lineage as well as macrophages, while other myeloerythroid cells, such as granulocytes, erythrocytes, and megakaryocytes could not be observed. Thus, IL7R$^+$c-Kit^{lo}Sca-1^{lo} cells harbor differentiating activity for the lymphoid as well as the myeloid lineage *in vitro*.

Differentiation of CD45$^+$CD4$^+$CD3$^-$ Cells Towards the Dendritic Cell Lineage
Since the fetal liver precursors could differentiate towards dendritic cells as well as CD45$^+$CD4$^+$CD3$^-$ cells *in vivo*, while we showed earlier that CD45$^+$CD4$^+$CD3$^-$ cells gave rise to antigen presenting cells *in vitro*, we asked whether CD45$^+$CD4$^+$CD3$^-$ cells were the only precursors to dendritic cells in developing lymph nodes. Hereto, newborn lymph nodes were divided into CD45$^+$CD4$^+$CD3$^-$ cells and CD45$^+$CD4$^-$ cells, and placed in a cytokine cocktail that allows for differentiation of lymphoid dendritic cells (Saunders et al., 1996). In these assays, we could only obtain professional antigen presenting cells from CD45$^+$CD4$^+$CD3$^-$ cells and not from the remaining lymph node cells.

Concluding Remarks
In fetal liver we have identified a cellular subset that is the phenotypic equivalent of the common lymphoid progenitor cells present in adult bone marrow. However, it appears that the developmental potential of these fetal liver precursors is less restricted than that of the adult bone marrow CLP, since the fetal liver cells can give rise to macrophages *in vitro*, while adult bone marrow CLP have lost this potential. It remains to be seen if the other progenies, CD45$^+$CD4$^+$CD3$^-$ cells and dendritic cells, derived from fetal liver cells can be derived from adult CLP.

The exact delineation of the precursors to CD45$^+$CD4$^+$CD3$^-$ cells will help us understand how certain genes are involved in lymphoid organ development. Genes might operate at the level of the fetal liver precursors when functional deletion of a gene affects both the development of lymphoid organs, as well as certain hematopoietic lineages (Georgopoulos et al., 1994; Yokota et al., 1999).

Thus, the fetal liver precursors differentiate into all lymphoid lineages, and are in this respect the fetal liver counterpart of common lymphoid progenitors. However, the fetal liver precursors have a broader potential since they can also give rise to dendritic cells and CD45$^+$CD4$^+$CD3$^-$ cells *in vivo*, as well as macrophages *in vitro*. We speculate that the mechanisms regulating this restricted potential of adult CLP can be found at the level of expressed transcription factors and cytokine receptors.

In conclusion we envision that the induction phase for lymph node and Peyer's patch development is a narrow window during embryonic development in which the attraction of fetal liver derived cells to certain predisposed niches formed by stromal cells leads to lymphoid organ induction and subsequent compartment formation within these organs. Only then can lymphocytes start populating the organs and adequately mount immune responses.

References

Alimzhanov, M. B., Kuprash, D. V., Kosco-Vilbois, M. H., Luz, A., Turetskaya, R. L., Tarakhovsky, A., Rajewsky, K., Nedospasov, S. A., and Pfeffer, K. (1997). Abnormal development of secondary lymphoid tissues in lymphotoxin beta-deficient mice. Proc. Natl. Acad. Sci. USA *94*, 9302-9307.

Banks, T. A., Rouse, B. T., Kerley, M. K., Blair, P. J., Godfrey, V. L., Kuklin, N. A., Bouley, D. M., Thomas, J., Kanangat, S., and Mucenski, M. L. (1995). Lymphotoxin-alpha-deficient mice. Effects on secondary lymphoid organ development and humoral immune responsiveness. J. Immunol. *155*, 1685-1693.

Burkly, L., Hession, C., Ogata, L., Reilly, C., Marconi, L. A., Olson, D., Tizard, R., Cate, R., and Lo, D. (1995). Expression of relB is required for the development of thymic medulla and dendritic cells. Nature *373*, 531-536.

Cao, X., Shores, E. W., Hu-Li, J., Anver, M. R., Kelsall, B. L., Russell, S. M., Drago, J., Noguchi, M., Grinberg, A., and Bloom, E. T. (1995). Defective lymphoid development in mice lacking expression of the common cytokine receptor gamma chain. Immunity *2*, 223-38.

Forster, R., Mattis, A. E., Kremmer, E., Wolf, E., Brem, G., and Lipp, M. (1996). A putative chemokine receptor, BLR1, directs B cell migration to defined lymphoid organs and specific anatomic compartments of the spleen. Cell *87*, 1037-1047.

Freeden-Jeffry, U. v., Vieira, P., Lucian, L. A., McNeil, T., Burdach, S. E., and Murray, R. (1995). Lymphopenia in interleukin (IL)-7 gene-deleted mice identifies IL-7 as a nonredundant cytokine. J. Exp. Med. *181*, 1519-1526.

Fu, Y. X., and Chaplin, D. D. (1999). Development and maturation of secondary lymphoid tissues. Annual Review Of Immunology *17*, 399-433.

Futterer, A., Mink, K., Luz, A., Kosco-Vilbois, M. H., and Pfeffer, K. (1998). The lymphotoxin beta receptor controls organogenesis and affinity maturation in peripheral lymphoid tissues. Immunity *9*, 59-70.

Georgopoulos, K., Bigby, M., Wang, J. H., Molnar, A., Wu, P., Winandy, S., and Sharpe, A. (1994). The Ikaros gene is required for the development of all lymphoid lineages. Cell *79*, 143-156.

Kaipainen, A., Korhonen, J., Mustonen, T., van Hinsbergh, V. W., Fang, G. H., Dumont, D., Breitman, M., and Alitalo, K. (1995). Expression of the fms-like tyrosine kinase 4 gene becomes restricted to lymphatic endothelium during development. Proceedings of the National Academy of Sciences of the United States of America *92*, 3566-70.

Kondo, M., Weissman, I. L., and Akashi, K. (1997). Identification of clonogenic common lymphoid progenitors in mouse bone marrow. Cell *91*, 661-672.

Kong, Y.-Y., Yoshida, H., Sarosi, I., Tan, H.-L., Timms, E., Capparelli, C., Morony, S., Oliveira-dos-Santos, A. J., Van, G., Itie, A., Koo, W., A.Wakeham, Dunstan, C. R., Lacey, D. L., Mak, T. W., Boyle, W. J., and Penninger, J. M. (1998). OPGL is a key regulator of osteoclastogenesis, lymphocyte development and lymph-node organogenesis. Nature *397*, 315-23.

Mebius, R. E., Rennert, P., and Weissman, I. L. (1997). Developing lymph nodes collect CD4+CD3-LTβ+ cells that can differentiate to APC, NK cells, and follicular cells but not T or B cells. Immunity *7*, 493-504.

Mebius, R. E., Streeter, P. R., Michie, S., Butcher, E. C., and Weissman, I. L. (1996). A developmental switch in lymphocyte homing receptor and endothelial vascular addressin expression regulates lymphocyte homing and permits CD4+ CD3- cells to colonize lymph nodes. Proc. Natl. Acad. Sci. USA *93*, 11019-24.

Rennert, P. D., Browning, J. L., Mebius, R., Mackay, F., and Hochman, P. S. (1996). Surface lymphotoxin alpha/beta complex is required for the development of peripheral lymphoid organs. J. Exp. Med. *184*, 1999-2006.

Rennert, P. D., James, D., Mackay, F., Browning, J. L., and Hochman, P. S. (1998). Lymph node genesis is induced by signaling through the lymphotoxin beta receptor. Immunity *9*, 71-79.

Sabin, F. R. (1909). The lymphatic system in human embryos, with a consideration of the morphology of the system as a whole. Am. J. Anat. *9*, 43-91.

Saunders, D., Lucas, K., Ismaili, J., Wu, L., Maraskovsky, E., Dunn, A., and Shortman, K. (1996). Dendritic cell development in culture from thymic precursor cells in the absence of granulocyte/macrophage colony-stimulating factor. J. Exp. Med. *184*, 2185-2196.

Togni, P. d., Goellner, J., Ruddle, N. H., Streeter, P. R., Fick, A., Mariathasan, S., Smith, S. C., Carlson, R., Shornick, L. P., Strauss-Schoenberger, J., Russell, J. H., Karr, R., and Chaplin, D. D. (1994). Abnormal development of peripheral lymphoid organs in mice deficient in lymphotoxin. Science *264*, 703-707.

Wang, J. H., Nichogiannopoulou, A., Wu, L., Sun, L., Sharpe, A. H., Bigby, M., and Georgopoulos, K. (1996). Selective defects in the development of the fetal and adult lymphoid system in mice with an Ikaros null mutation. Immunity *5*, 537-549.

Weih, F., Carrasco, D., Durham, S. K., Barton, D. S., Rizzo, C. A., Ryseck, R. P., Lira, S. A., and Bravo, R. (1995). Multiorgan inflammation and hematopoietic abnormalities in mice with a targeted disruption of RelB, a member of the NF-kappa B/Rel family. Cell *80*, 331-340.

Wigle, J. T., and Oliver, G. (1999). Prox1 function is required for the development of the murine lymphatic system. Cell *98*, 769-778.

Yokota, Y., Mansouri, A., Mori, S., Sugawara, S., Adachi, S., Nishikawa, S.-I., and Gruss, P. (1999). Development of peripheral lymphoid organs and natural killer cells depends on the helix-loop-helix inhibitor Id2. Nature *397*, 702-706.

[illegible]
[illegible]
[illegible]
[illegible]
[illegible]
[illegible]
Aguzzi, A. [illegible] (1999). [illegible] for the development of [illegible] [illegible]
Klein, M. A., [illegible], Mack, S., [illegible] (1997). A crucial role for B cells in neuroinvasive scrapie. Nature 390, 687-690.

The Role of Tumor Necrosis Factor and Lymphotoxin in Lymphoid Organ Development

R. Ettinger
Basel Institute for Immunology, Postfach, 4005 Basel, Switzerland

Tumor necrosis factor (TNF), lymphotoxin-α (LTα) and LTβ are related cytokines which belong to the TNF ligand family and are encoded by genes clustered within the MHC gene complex [1, 2]. TNF and LTα self-associate as homotrimers, capable of binding both the 55-60 KD TNF receptor 1 (TNFRp55) and the 75-80 KD TNF receptor 2 (TNFRp75) [3]. In addition, LTα also binds herpes simplex virus (HSV) entry mediator (HVEM) which is an orphan member of the TNF receptor family [4-6]. LTβ exists in a heterotrimeric complex with LTα, which solely binds the lymphotoxin-β receptor (LTβR) [1, 7-10]. Recently, a novel ligand which shares homology with TNF family members has been found and termed LIGHT [4]. Like the other ligands, LIGHT forms a homotrimer and is capable of binding two receptors, HVEM and LTβR [4] (see Fig. 1).

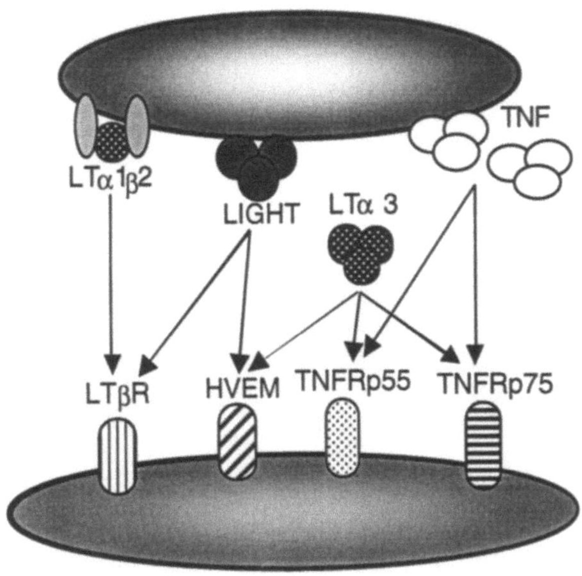

Fig. 1 TNF, LT & LIGHT ligands and their receptors.
Cell bound and soluble TNF and soluble LTα homotrimers bind both TNFRp55 and TNFRp75. Soluble LTα3 also binds to HVEM. Membrane-bound LTα1β2 is only known to bind LTβR, but not the TNFRp55, TNFRp75 or HVEM molecules. LIGHT has been shown to bind both HVEM and LTβR.

Both TNF and LT have been extensively characterized and have been demonstrated to induce a variety of physiological effects [11]. However, only recently has it been discovered that these cytokines are required for the normal development of the secondary lymphoid organs. We and others have previously demonstrated that LTα1β2/LTβR and TNF/TNFRp55 are crucial interactions for lymphoid organogenesis. Disruption of signaling through either the LTβR or the TNFRp55 by selective gene targeting (of receptors or ligands) or expression of a neutralizing transgene fusion protein, results in altered peripheral lymphoid organ development (see Table 1) [12-23]. Although distinct phenotypes, mice devoid of LTβR or TNFRp55 signaling develop disrupted splenic architecture, lacking primary follicles, follicular dendritic cells (FDC) and germinal centers, but interdigitating dendritic cells (IDC) appear not to be affected (Fig 2). In LTα–, LTβ– or LTβR-deficient mice, or mice expressing an LTβR-Fc fusion protein, the splenic marginal zones are absent [12, 16, 18, 22, 24]. In contrast, splenic marginal zones do develop in TNF-, or TNFRp55-deficient mice, or mice expressing a TNFRp55-Fc fusion protein [13, 15]. Moreover, marginal zones are enlarged in mice expressing the TNFRp55-Fc chimeric protein. Lymph node and Peyer's patch development is also affected in many of these animals [12-14, 16, 18, 19, 22-28]. Interestingly, germinal centers are present in the absence of FDC networks in lymph nodes and Peyer's patches [12, 29, 30]. These data demonstrate that FDC are not required for germinal center formation. The roles of LIGHT and HVEM in lymphoid organogenesis have not yet been evaluated.

Table 1: TNF and LT are required for normal lymphoid organogenesis

	TNF-/-	TNFR p55-/-	TNFR p75-/-	LTα-/-	LTβ-/-	LTβR-/-
Thymus	+	+	+	+	+	+
Spleen	+	+	+	+	+	+
1' follicles	-	-	+	-	+/-	-
FDC	-	-	+	-	-	-
GC	-	-	+	-	-	-
MZ	+	+*	+	-	-	-
Lymph nodes	+	+	+	-*	M & C¶	-
1' follicles	-	-	+	-	+	-
FDC	-	-	+	-	-	-
GC	nd§	+	+	-	+	-
Peyer's Patches	+	+*	+	-	-	-

*some discrepancies exist among independently generated mouse lines

¶mesenteric & cervical

§not determined

no data on LIGHT and HVEM

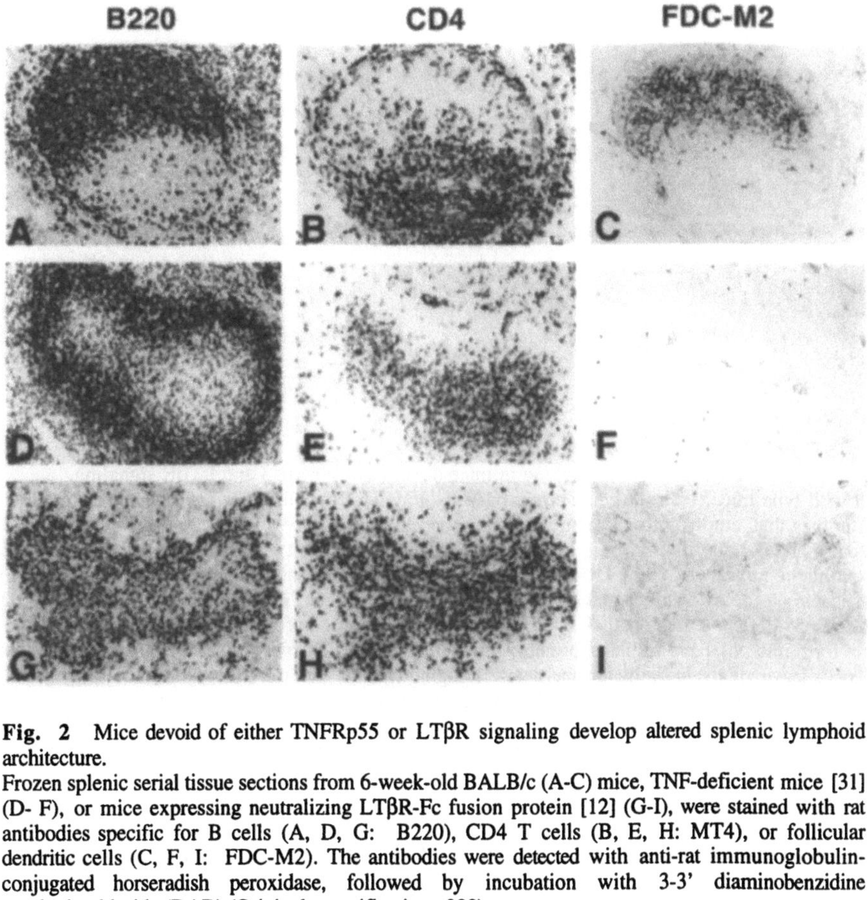

Fig. 2 Mice devoid of either TNFRp55 or LTβR signaling develop altered splenic lymphoid architecture.
Frozen splenic serial tissue sections from 6-week-old BALB/c (A-C) mice, TNF-deficient mice [31] (D- F), or mice expressing neutralizing LTβR-Fc fusion protein [12] (G-I), were stained with rat antibodies specific for B cells (A, D, G: B220), CD4 T cells (B, E, H: MT4), or follicular dendritic cells (C, F, I: FDC-M2). The antibodies were detected with anti-rat immunoglobulin-conjugated horseradish peroxidase, followed by incubation with 3-3' diaminobenzidine tetrahydrochloride (DAB) (Original magnification x200).

Lymphocytes have been shown to migrate into the spleen via terminal arterioles which branch into both the marginal zone and marginal sinus. B and T cells then enter into the white pulp and first traffic through the inner periarteriolar lymphoid sheath (PALS), where T cells interact with IDC. The B cells then migrate into follicles which form around FDC networks [32, 33]. In TNF-deficient or TNFRp55-deficient mice, B cells presumably still traffic through the inner T cell areas before entering the B cell zones. However, the lack of FDC networks likely prevents B cell polarization and follicle formation, resulting in a ring-like structure of B cells at the outer PALS. A distinct phenotype occurs in mice deficient in LTβ, LTα, or LTβR expression. In these mice, the B cells enter the T cell areas, but are unable to gain entry into the B cell zone. This results in B cells, T cells and IDC all contained within the inner PALS and it appears that the lack of splenic FDC networks is due to the inability of B cells to enter into the follicles (Fig. 3).

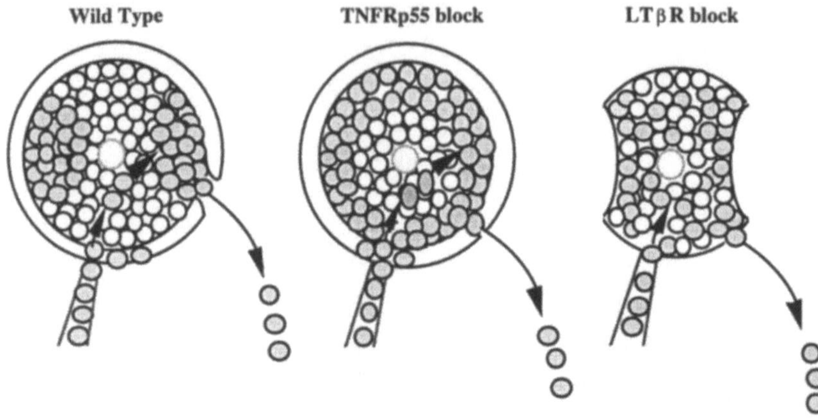

Fig. 3 Possible lymphocyte trafficking in mice devoid of TNFRp55 and LTβR signaling. In wild type mice, both T (white circles) and B (gray circles) cells enter the spleen via terminal arterioles that empty into the marginal sinus and marginal zone (surrounding the white pulp). Both T and B cells first traffic into the inner PALS, where T cells intertwine with IDC. B cells continue to migrate into the FDC-containing B cell-rich follicles, and either form germinal centers, or leave via the bridging channels (breaks in the marginal zone) into the red pulp. In TNF- or TNFRp55-deficient mice, both T and B cells appear to migrate properly, however, the B cells do not form into polarized follicles, possibly due to the lack of FDC networks. It is not known if the B cells leave via the bridging channels since bridging channels are not readily visible in these mice. In mice deficient for LTα, LTβ or LTβR expression, both T and B cells appear to migrate into the inner PALS. However, the B cells do not enter (or form) B cell follicles, nor do FDC develop. In these mice there is no marginal zone, thus it is not clear how the cells enter or leave the white pulp areas.

It is not clear what initiates the formation of B cell follicles and FDC development. Presumably, a specialized B cell expressing high levels of TNF and LT enters the spleen early during development. This cell may activate the stroma to differentiate into FDC, which the entering B cells then bind, forming polarized follicles. Other specialized cells may also be involved in the early formation of the splenic microenvironment, such as $CD4^{+}CD3^{-}$ cells which are known to express high levels of LT [34]. It has previously been demonstrated that hematopoietic cells are essential for the development of FDC networks in SCID mice reconstituted with selective lymphocyte populations [35-37]. More recently, it has been shown that TNF-, LTα– and LTβ–expressing B cells and TNFRp55- and LTβR-expressing radioresistant stromal cells are required for the normal development of splenic architecture [38-43].

The exact mechanism of TNF and/or LT actions on lymphoid organ development is unknown, however, chemokines have been demonstrated to play an important role. Secondary lymphoid organ chemokine (SLC) expressed on high endothelial venules (HEV) and on stromal cells of lymphoid T cell areas, EBV-induced molecule 1 ligand chemokine (ELC) expressed by a subset of dendritic cells (DC), and B lymphocyte chemoattractant (BLC) expressed on stromal cells in B cell follicles, have all been shown to be important in lymphocyte migration [44-49]. In addition, mice devoid of the BLC receptor, Burkitt's lymphoma receptor-1 (BLR-1), also have defects in B cell trafficking [50-52]. BLR-1-deficient mice develop phenotypes similar to those observed in TNF- and TNFRp55-deficient animals, lacking splenic and Peyer's patch B cell follicles [52]. Moreover, BLC, SLC and ELC have been shown to be reduced in TNF-, TNFRp55-, LTα- and LTβ-deficient mice, and it was suggested that TNF and LT are required for the normal expression of these chemokines [53]. Thus, it appears that the disrupted lymphoid architecture in TNF- and LT-null mice is due to a reduction/lack of specific chemokines, whose expression is dependent on the presence of TNF and LT molecules. However, the reduction of SLC and BLC message may be secondary to the loss of a specific cell type, not the lack of TNF or LT expression. Pfeffer and colleagues have demonstrated that although they do not detect SLC or BLC message in the spleens of LTβR-deficient mice, they do detect these chemokines in the lungs (which become highly infiltrated with CD4+ T cells and B cells) by Northern blot analysis (K. Peffer, personal communication, [22]). These data suggest that SLC and BLC expression is not dependent on LT, but that the spleens of LTα– and LTβ–deficient mice lack a critical cell population which is required for the upregulation of these chemokines. Presumably, this population induces BLC (and SLC) expression, which is then followed by the migration of B cells into the follicular areas of the spleen, and formation of FDC networks.

While it is clear that TNF and LT control peripheral lymphoid organ development, many questions remain with respect to molecular mechanisms. Future studies need to address which specific hematopoietic cell subpopulations initiate critical events in lymphoid organogenesis. Moreover, the reticular stromal cells that respond to these signals must be identified. This fundamental information will give new insights into the development and maintenance of lymphoid tissue architecture.

Acknowledgements

We would like to thank Leslie Nicklin, Lucy Trippmacher, Allison Dwileski and Beatrice Pfeiffer for assistance with the manuscript, and Paul Kincade and Marina Cella for critical reading of the manuscript. We also thank Marie Kosco-Vilbois for providing the anti-FDC (FDC-M2 and FDC-M1) monoclonal antibodies. The Basel Institute for Immunology was founded and is supported by F. Hoffmann-La Roche Ltd.

References

1. Browning JL, Ngam-ek A, Lawton P, et al. (1993) Lymphotoxin beta, a novel member of the TNF family that forms a heteromeric complex with lymphotoxin on the cell surface. Cell 72:847-856
2. Lawton P, Nelson J, Tizard R, Browning JL (1995) Characterization of the mouse lymphotoxin-beta gene. J Immunol 154:239-246
3. Tartaglia LA, Goeddel DV (1992) Two TNF receptors. Immunol Today 13:151-153
4. Mauri DN, Ebner R, Montgomery RI, et al. (1998) LIGHT, a new member of the TNF superfamily, and lymphotoxin alpha are ligands for herpesvirus entry mediator. Immunity 8:21-30
5. Ware CF, VanArsdale TL, Crowe PD, Browning JL (1995) The ligands and receptors of the lymphotoxin system. In: Griffiths GM, Tschopp J, eds. Curr. Top. Microbiol. Immunol. Basel: Springer-Verlag, 175-218. vol 198

6. Smith CA, Farrah T, Goodwin RG (1994) The TNF receptor superfamily of cellular and viral proteins: activation, costimulation, and death. Cell 76:959-962
7. Androlewicz MJ, Browning JL, Ware CF (1992) Lymphotoxin is expressed as a heteromeric complex with a distinct 33- kDa glycoprotein on the surface of an activated human T cell hybridoma. J Biol Chem 267:2542-2547
8. Browning JL, Miatkowski K, Griffiths DA, et al. (1996) Preparation and characterization of soluble recombinant heterotrimeric complexes of human lymphotoxins alpha and beta. J Biol Chem 271:8618-8626
9. Crowe PD, VanArsdale TL, Walter BN, et al. (1994) A lymphotoxin-beta-specific receptor. Science 264:707-710
10. Force WR, Walter BN, Hession C, et al. (1995) Mouse lymphotoxin-beta receptor. Molecular genetics, ligand binding, and expression. J Immunol 155:5280-5288
11. Vassalli P (1992 The pathophysiology of tumor necrosis factors. Annu Rev Immunol 10:411-452
12. Ettinger R, Browning JL, Michie SA, van Ewijk W, McDevitt HO (1996) Disrupted splenic architecture, but normal lymph node development in mice expressing a soluble lymphotoxin-beta receptor-IgG1 fusion protein. Proc Natl Acad Sci USA 93:13102-13107
13. Ettinger R, Mebius R, Browning JL, et al. (1998) Effects of tumor necrosis factor and lymphotoxin on peripheral lymphoid tissue development. Int Immunol 10:727-741
14. Banks TA, Rouse BT, Kerley MK, et al. (1995) Lymphotoxin-alpha-deficient mice. Effects on secondary lymphoid organ development and humoral immune responsiveness. J Immunol 155:1685-1693
15. Pasparakis M, Alexopoulou L, Episkopou V, Kollias G (1996) Immune and inflammatory responses in TNF alpha-deficient mice: a critical requirement for TNF alpha in the formation of primary B cell follicles, follicular dendritic cell networks and germinal centers, and in the maturation of the humoral immune response. J Exp Med 184:1397-1411
16. Rennert PD, Browning JL, Mebius R, Mackay F, Hochman PS (1996) Surface lymphotoxin alpha/beta complex is required for the development of peripheral lymphoid organs. J Exp Med 184:1999-2006
17. Le Hir M, Bluethmann H, Kosco-Vilbois MH, et al. (1996) Differentiation of follicular dendritic cells and full antibody responses require tumor necrosis factor receptor-1 signaling. J Exp Med 183:2367-2372
18. Koni PA, Sacca R, Lawton P, Browning JL, Ruddle NH, Flavell RA (1997) Distinct roles in lymphoid organogenesis for lymphotoxins alpha and beta revealed in lymphotoxin beta-deficient mice. Immunity 6:491-500
19. Neumann B, Luz A, Pfeffer K, Holzmann B (1996) Defective Peyer's patch organogenesis in mice lacking the 55-kD receptor for tumor necrosis factor. J Exp Med 184:259-264
20. de Kossodo S, Grau GE, Daneva T, et al. (1992) Tumor necrosis factor alpha is involved in mouse growth and lymphoid tissue development. J Exp Med 176:1259-1264
21. Matsumoto M, Mariathasan S, Nahm MH, Baranyay F, Peschon JJ, Chaplin DD (1996) Role of lymphotoxin and the type I TNF receptor in the formation of germinal centers. Science 271:1289-1291
22. Futterer A, Mink K, Luz A, Kosco-Vilbois MH, Pfeffer K (1998) The lymphotoxin beta receptor controls organogenesis and affinity maturation in peripheral lymphoid tissues. Immunity 9:59-70
23. Korner H, Cook M, Riminton DS, et al. (1997) Distinct roles for lymphotoxin-alpha and tumor necrosis factor in organogenesis and spatial organization of lymphoid tissue. Eur J Immunol 27:2600-2609
24. De Togni P, Goellner J, Ruddle NH, et al. (1994) Abnormal development of peripheral lymphoid organs in mice deficient in lymphotoxin. Science 264:703-707
25. Alimzhanov MB, Kuprash DV, Kosco-Vilbois MH, et al. (1997) Abnormal development of secondary lymphoid tissues in lymphotoxin beta- deficient mice. Proc Natl Acad Sci USA 94:9302-9307

26. Pasparakis M, Alexopoulou L, Grell M, Pfizenmaier K, Bluethmann H, Kollias G (1997) Peyer's patch organogenesis is intact yet formation of B lymphocyte follicles is defective in peripheral lymphoid organs of mice deficient for tumor necrosis factor and its 55-kDa receptor [published erratum appears in Proc Natl Acad Sci U S A (1997) Aug 19;94(17):9510]. Proc Natl Acad Sci USA 94:6319-6323
27. Rennert PD, Browning JL, Hochman PS (1997) Selective disruption of lymphotoxin ligands reveals a novel set of mucosal lymph nodes and unique effects on lymph node cellular organization. Int Immunol 9:1627-1639
28. Eugster HP, Muller M, Karrer U, et al. (1996) Multiple immune abnormalities in tumor necrosis factor and lymphotoxin- alpha double-deficient mice. Int Immunol 8:23-36
29. Koni PA, Flavell RA (1999) Lymph node germinal centers form in the absence of follicular dendritic cell networks. J Exp Med 189:855-864
30. Fu YX, Huang G, Matsumoto M, Molina H, Chaplin DD (1997) Independent signals regulate development of primary and secondary follicle structure in spleen and mesenteric lymph node. Proc Natl Acad Sci U S A 94:5739-5743
31. Marino MW, Dunn A, Grail D, et al. (1997) Characterization of tumor necrosis factor-deficient mice. Proc Natl Acad Sci USA 94:8093-8098
32. van Ewijk W, Nieuwenhuis P (1985) Compartments, domains and migration pathways of lymphoid cells in the splenic pulp. Experientia 41:199-208
33. Goodnow CC, Cyster JG (1997) Lymphocyte homing: the scent of a follicle. Curr Biol 7:R219-222
34. Mebius RE, Rennert P, Weissman IL (1997) Developing lymph nodes collect CD4+CD3- LTbeta+ cells that can differentiate to APC, NK cells, and follicular cells but not T or B cells. Immunity 7:493-504
35. Kapasi ZF, Burton GF, Shultz LD, Tew JG, Szakal AK (1993) Induction of functional follicular dendritic cell development in severe combined immunodeficiency mice. Influence of B and T cells. J Immunol 150:2648-2658
36. Yoshida K, Matsuura N, Tamahashi N, Takahashi T (1993) Development of antigenic heterogeneity in the splenic meshwork of severe combined immunodeficient (SCID) mice after reconstitution with T and B lymphocytes. Cell Tissue Res 272:1-10
37. Yoshida K, van den Berg TK, Dijkstra CD (1994) The functional state of follicular dendritic cells in severe combined immunodeficient (SCID) mice: role of the lymphocytes. Eur J Immunol 24:464-468
38. Endres R, Alimzhanov MB, Plitz T, et al. (1999) Mature follicular dendritic cell networks depend on expression of lymphotoxin beta receptor by radioresistant stromal cells and of lymphotoxin beta and tumor necrosis factor by B cells. J Exp Med 189:159-168
39. Fu YX, Huang G, Wang Y, Chaplin DD (1998) B lymphocytes induce the formation of follicular dendritic cell clusters in a lymphotoxin alpha-dependent fashion. J Exp Med 187:1009-1018
40. Gonzalez M, Mackay F, Browning JL, Kosco-Vilbois MH, Noelle RJ (1998) The sequential role of lymphotoxin and B cells in the development of splenic follicles. J Exp Med 187:997-1007
41. Matsumoto M, Fu YX, Molina H, et al. (1997) Distinct roles of lymphotoxin alpha and the type I tumor necrosis factor (TNF) receptor in the establishment of follicular dendritic cells from non-bone marrow-derived cells. J Exp Med 186:1997-2004
42. Tkachuk M, Bolliger S, Ryffel B, et al. (1998) Crucial role of tumor necrosis factor receptor 1 expression on nonhematopoietic cells for B cell localization within the splenic white pulp. J Exp Med 187:469-477
43. Cook MC, Korner H, Riminton DS, et al. (1998) Generation of splenic follicular structure and B cell movement in tumor necrosis factor-deficient mice. J Exp Med 188:1503-1510
44. Gunn MD, Tangemann K, Tam C, Cyster JG, Rosen SD, Williams LT (1998) A chemokine expressed in lymphoid high endothelial venules promotes the adhesion and chemotaxis of naive T lymphocytes. Proc Natl Acad Sci USA 95:258-263

45. Tanabe S, Lu Z, Luo Y, et al. (1997) Identification of a new mouse beta-chemokine, thymus-derived chemotactic agent 4, with activity on T lymphocytes and mesangial cells. J Immunol 159:5671-5679
46. Ngo VN, Tang HL, Cyster JG (1998) Epstein-Barr virus-induced molecule 1 ligand chemokine is expressed by dendritic cells in lymphoid tissues and strongly attracts naive T cells and activated B cells. J Exp Med 188:181-191
47. Gunn MD, Ngo VN, Ansel KM, Ekland EH, Cyster JG, Williams LT (1998) A B-cell-homing chemokine made in lymphoid follicles activates Burkitt's lymphoma receptor-1. Nature 391:799-803
48. Legler DF, Loetscher M, Roos RS, Clark-Lewis I, Baggiolini M, Moser B (1998) B cell-attracting chemokine 1, a human CXC chemokine expressed in lymphoid tissues, selectively attracts B lymphocytes via BLR1/CXCR5. J Exp Med 187:655-660
49. Gunn MD, Kyuwa S, Tam C, et al. (1999) Mice lacking expression of secondary lymphoid organ chemokine have defects in lymphocyte homing and dendritic cell localization. J Exp Med 189:451-460
50. Dobner T, Wolf I, Emrich T, Lipp M (1992) Differentiation-specific expression of a novel G protein-coupled receptor from Burkitt's lymphoma. Eur J Immunol 22:2795-2799
51. Kaiser E, Forster R, Wolf I, Ebensperger C, Kuehl WM, Lipp M (1993) The G protein-coupled receptor BLR1 is involved in murine B cell differentiation and is also expressed in neuronal tissues. Eur J Immunol 23:2532-2539
52. Forster R, Mattis AE, Kremmer E, Wolf E, Brem G, Lipp M (1996) A putative chemokine receptor, BLR1, directs B cell migration to defined lymphoid organs and specific anatomic compartments of the spleen. Cell 87:1037-1047
53. Ngo VN, Korner H, Gunn MD, et al. (1999) Lymphotoxin alpha/beta and tumor necrosis factor are required for stromal cell expression of homing chemokines in B and T cell areas of the spleen. J Exp Med 189:403-412

Preface

Current Topics in Microbiology and Immunology

Volumes published since 1989 (and still available)

Vol. 210: **Potter, Michael; Rose, Noel R. (Eds.):** Immunology of Silicones. 1996. 136 figs. XX, 430 pp. ISBN 3-540-60272-0

Vol. 211: **Wolff, Linda; Perkins, Archibald S. (Eds.):** Molecular Aspects of Myeloid Stem Cell Development. 1996. 98 figs. XIV, 298 pp. ISBN 3-540-60414-6

Vol. 212: **Vainio, Olli; Imhof, Beat A. (Eds.):** Immunology and Developmental Biology of the Chicken. 1996. 43 figs. IX, 281 pp. ISBN 3-540-60585-1

Vol. 213/I: **Günthert, Ursula; Birchmeier, Walter (Eds.):** Attempts to Understand Metastasis Formation I. 1996. 35 figs. XV, 293 pp. ISBN 3-540-60680-7

Vol. 213/II: **Günthert, Ursula; Birchmeier, Walter (Eds.):** Attempts to Understand Metastasis Formation II. 1996. 33 figs. XV, 288 pp. ISBN 3-540-60681-5

Vol. 213/III: **Günthert, Ursula; Schlag, Peter M.; Birchmeier, Walter (Eds.):** Attempts to Understand Metastasis Formation III. 1996. 14 figs. XV, 262 pp.
ISBN 3-540-60682-3

Vol. 214: **Kräusslich, Hans-Georg (Ed.):** Morphogenesis and Maturation of Retroviruses. 1996. 34 figs. XI, 344 pp.
ISBN 3-540-60928-8

Vol. 215: **Shinnick, Thomas M. (Ed.):** Tuberculosis. 1996. 46 figs. XI, 307 pp.
ISBN 3-540-60985-7

Vol. 216: **Rietschel, Ernst Th.; Wagner, Hermann (Eds.):** Pathology of Septic Shock. 1996. 34 figs. X, 321 pp.
ISBN 3-540-61026-X

Vol. 217: **Jessberger, Rolf; Lieber, Michael R. (Eds.):** Molecular Analysis of DNA Rearrangements in the Immune System. 1996. 43 figs. IX, 224 pp. ISBN 3-540-61037-5

Vol. 218: **Berns, Kenneth I.; Giraud, Catherine (Eds.):** Adeno-Associated Virus (AAV) Vectors in Gene Therapy. 1996. 38 figs. IX,173 pp. ISBN 3-540-61076-6

Vol. 219: **Gross, Uwe (Ed.):** Toxoplasma gondii. 1996. 31 figs. XI, 274 pp.
ISBN 3-540-61300-5

Vol. 220: **Rauscher, Frank J. III; Vogt, Peter K. (Eds.):** Chromosomal Translocations and Oncogenic Transcription Factors. 1997. 28 figs. XI, 166 pp. ISBN 3-540-61402-8

Vol. 221: **Kastan, Michael B. (Ed.):** Genetic Instability and Tumorigenesis. 1997. 12 figs.VII, 180 pp.
ISBN 3-540-61518-0

Vol. 222: **Olding, Lars B. (Ed.):** Reproductive Immunology. 1997. 17 figs. XII, 219 pp.
ISBN 3-540-61888-0

Vol. 223: **Tracy, S.; Chapman, N. M.; Mahy, B. W. J. (Eds.):** The Coxsackie B Viruses. 1997. 37 figs. VIII, 336 pp.
ISBN 3-540-62390-6

Vol. 224: **Potter, Michael; Melchers, Fritz (Eds.):** C-Myc in B-Cell Neoplasia. 1997. 94 figs. XII, 291 pp. ISBN 3-540-62892-4

Vol. 225: **Vogt, Peter K.; Mahan, Michael J. (Eds.):** Bacterial Infection: Close Encounters at the Host Pathogen Interface. 1998. 15 figs. IX, 169 pp.
ISBN 3-540-63260-3

Vol. 226: **Koprowski, Hilary; Weiner, David B. (Eds.):** DNA Vaccination/Genetic Vaccination. 1998. 31 figs. XVIII, 198 pp.
ISBN 3-540-63392-8

Vol. 227: **Vogt, Peter K.; Reed, Steven I. (Eds.):** Cyclin Dependent Kinase (CDK) Inhibitors. 1998. 15 figs. XII, 169 pp.
ISBN 3-540-63429-0

Vol. 228: **Pawson, Anthony I. (Ed.):** Protein Modules in Signal Transduction. 1998. 42 figs. IX, 368 pp. ISBN 3-540-63396-0

Vol. 229: **Kelsoe, Garnett; Flajnik, Martin (Eds.):** Somatic Diversification of Immune Responses. 1998. 38 figs. IX, 221 pp.
ISBN 3-540-63608-0

Vol. 230: **Kärre, Klas; Colonna, Marco (Eds.):** Specificity, Function, and Development of NK Cells. 1998. 22 figs. IX, 248 pp. ISBN 3-540-63941-1

Vol. 231: **Holzmann, Bernhard; Wagner, Hermann (Eds.):** Leukocyte Integrins in the Immune System and Malignant Disease. 1998. 40 figs. XIII, 189 pp. ISBN 3-540-63609-9

Vol. 232: **Whitton, J. Lindsay (Ed.):** Antigen Presentation. 1998. 11 figs. IX, 244 pp. ISBN 3-540-63813-X

Vol. 233/I: **Tyler, Kenneth L.; Oldstone, Michael B. A. (Eds.):** Reoviruses I. 1998. 29 figs. XVIII, 223 pp. ISBN 3-540-63946-2

Vol. 233/II: **Tyler, Kenneth L.; Oldstone, Michael B. A. (Eds.):** Reoviruses II. 1998. 45 figs. XVI, 187 pp. ISBN 3-540-63947-0

Vol. 234: **Frankel, Arthur E. (Ed.):** Clinical Applications of Immunotoxins. 1999. 16 figs. IX, 122 pp. ISBN 3-540-64097-5

Vol. 235: **Klenk, Hans-Dieter (Ed.):** Marburg and Ebola Viruses. 1999. 34 figs. XI, 225 pp. ISBN 3-540-64729-5

Vol. 236: **Kraehenbuhl, Jean-Pierre; Neutra, Marian R. (Eds.):** Defense of Mucosal Surfaces: Pathogenesis, Immunity and Vaccines. 1999. 30 figs. IX, 296 pp. ISBN 3-540-64730-9

Vol. 237: **Claesson-Welsh, Lena (Ed.):** Vascular Growth Factors and Angiogenesis. 1999. 36 figs. X, 189 pp. ISBN 3-540-64731-7

Vol. 238: **Coffman, Robert L.; Romagnani, Sergio (Eds.):** Redirection of Th1 and Th2 Responses. 1999. 6 figs. IX, 148 pp. ISBN 3-540-65048-2

Vol. 239: **Vogt, Peter K.; Jackson, Andrew O. (Eds.):** Satellites and Defective Viral RNAs. 1999. 39 figs. XVI, 179 pp. ISBN 3-540-65049-0

Vol. 240: **Hammond, John; McGarvey, Peter; Yusibov, Vidadi (Eds.):** Plant Biotechnology. 1999. 12 figs. XII, 196 pp. ISBN 3-540-65104-7

Vol. 241: **Westblom, Tore U.; Czinn, Steven J.; Nedrud, John G. (Eds.):** Gastroduodenal Disease and Helicobacter pylori. 1999. 35 figs. XI, 313 pp. ISBN 3-540-65084-9

Vol. 242: **Hagedorn, Curt H.; Rice, Charles M. (Eds.):** The Hepatitis C Viruses. 2000. 47 figs. IX, 379 pp. ISBN 3-540-65358-9

Vol. 243: **Famulok, Michael; Winnacker, Ernst-L.; Wong, Chi-Huey (Eds.):** Combinatorial Chemistry in Biology. 1999. 48 figs. IX, 189 pp. ISBN 3-540-65704-5

Vol. 244: **Daëron, Marc; Vivier, Eric (Eds.):** Immunoreceptor Tyrosine-Based Inhibition Motifs. 1999. 20 figs. VIII, 179 pp. ISBN 3-540-65789-4

Vol. 245/I: **Justement, Louis B.; Siminovitch, Katherine A. (Eds.):** Signal Transduction and the Coordination of B Lymphocyte Development and Function I. 2000. 22 figs. XVI, 274 pp. ISBN 3-540-66002-X

Vol. 245/II: **Justement, Louis B.; Siminovitch, Katherine A. (Eds.):** Signal Transduction on the Coordination of B Lymphocyte Development and Function II. 2000. 13 figs. XV, 172 pp. ISBN 3-540-66003-8

Vol. 246: **Melchers, Fritz; Potter, Michael (Eds.):** Mechanisms of B Cell Neoplasia 1998. 1999. 111 figs. XXIX, 415 pp. ISBN 3-540-65759-2

Vol. 247: **Wagner, Hermann (Ed.):** Immunobiology of Bacterial CpG-DNA. 2000. 34 figs. IX, 246 pp. ISBN 3-540-66400-9

Vol. 248: **du Pasquier, Louis; Litman, Gary W. (Eds.):** Origin and Evolution of the Vertebrate Immune System. 2000. 81 figs. IX, 324 pp. ISBN 3-540-66414-9

Vol. 249: **Jones, Peter A.; Vogt, Peter K. (Eds.):** DNA Methylation and Cancer. 2000. 16 figs. IX, 169 pp. ISBN 3-540-66608-7

Vol. 250: **Aktories, Klaus; Wilkins, Tracy, D. (Eds.):** Clostridium difficile. 2000. 20 figs. IX, 143 pp. ISBN 3-540-67291-5

MIX
Papier aus verantwortungsvollen Quellen
Paper from responsible sources
FSC® C105338

If you have any concerns about our products,
you can contact us on
ProductSafety@springernature.com

In case Publisher is established outside the EU,
the EU authorized representative is:
Springer Nature Customer Service Center GmbH
Europaplatz 3, 69115 Heidelberg, Germany

Printed by Libri Plureos GmbH
in Hamburg, Germany